Bliss Bibliographic Classification

Bliss Bibliographic Classification

Second Edition

J. Mills and Vanda Broughton

School of Information and Communication Studies
University of North London

Class AY General Science
Class B Physics

BOWKER SAUR

LONDON MELBOURNE MUNICH NEW PROVIDENCE, NEW JERSEY

First edition by Henry Evelyn Bliss published under the title
A Bibliographic Classification.
Volume 1 – 4, 1940 – 1953

Second edition by J. Mills and Vanda Broughton published under the title
Bliss Bibliographic Classification. 1977–.

British Library Cataloguing in Publication Data

A catalogue record for this book is available from the British Library

Library of Congress Cataloging in Publication Data

A catalog record for this book is available from the Library of Congress

Published by Bowker-Saur
Tel: +44 (0) 1342 326972 Fax: +44 (0) 1342 335612
E-mail: lis@bowker-saur.co.uk
Web site: www.bowker-saur.co.uk

Bowker-Saur is a division of REED BUSINESS INFORMATION

ISBN 0408708247

Printed on acid-free paper

Printed by Antony Rowe Ltd, Chippenham, Wiltshire

Contents

Outline of the complete BC2 classification

2	Physical forms and forms of arrangement of documents
3	Phenomena: objects of knowledge
	Prolegomena
5	Knowledge, information
	Disciplines: forms of knowledge
A	Philosophy
AM	Mathematics
AY	Science
B	Physical sciences
E	Biological sciences
H	Human sciences and human studies
	Physical anthropology, health and medicine, psychology
J/T	Social sciences and humanities
U/V	Technology and useful arts
W/Y	The Arts (including music and philology)

Second outline

2	Generalia: physical forms & forms of arrangement of documents
3	Phenomena: objects of knowledge
	* For completely multidisciplinary documents, unclassifiable under any one discipline.
	Attributes, activities & processes, entities
	Prolegomena
6	Universe of knowledge
	Methods of enquiry... Disciplines (general)
7	Information science & technology
	Information processing... computers
8	Information work... retrieval
	Information systems & services
	Primarily reference retrieval... computerised services
9	Primarily document retrieval... libraries
	Disciplines: forms of knowledge
A	Philosophy & logic
AM	Mathematics, statistics & probability
AY	Sciences, natural sciences
B	Physics
	Physics-based technologies
	* Alternative only.
C	Chemistry
	Chemistry-based technologies
	* Alternative only.
D	Astronomy & space sciences
DH	Earth sciences
	Geophysics, geology, hydrology, meteorology...
	Geography: regional... systematic...
E	Biological sciences
EK	Microorganisms
F	Botany
G	Zoology
GR	Applied biology
	Agriculture & animal exploitation
GY	General & human ecology
H	Human sciences & human studies
HA	Human biology... physical anthropology
HH	Health & medicine
I	Psychology
J	Education
K	Society
K9Q	Perspectives: sociology, social anthropology...
KAH	Social ecology & environment... demography...
KC	Social processes: change... social behaviour...
KK	Social structure
KLK	Collectivities: groups... classes... family...
KRS	Inclusive societies... non-literate... literate...
KW	Customs, folklore & mythology
LA	Area studies
LB	Geography
	* Alternative only.
LC	Travel & description, topography
LD	History
LE	Study of history... auxiliary sciences: archaeology
LF	Prehistory
LG	*By social activity*
	Social history... political history...
	* Alternative only.
	[History LD]
	By ethnic group
	By broad period
LI	Ancient history
	Modern history
M	Favoured country
	* For example, UK in British libraries.
N	Other countries
	As Schedule 2.
O	Local history... biography...
	* Alternative only for local history.
P	Religion... occult... morals & ethics
Q	Social welfare & criminology
R	Politics... public administration
S	Law
T	Economics... management of enterprises
U/V	Technology & useful arts
	Equipment, plant, instrumentation
	Systems engineering... control... computers...
	Technical testing, maintenance, design... production technology...
	Materials handling... packaging... storing...
	Energy technologies (general)
	Physics–based technologies (nuclear, electrical, thermal, mechanical technologies)
	Construction technology... architecture... physical planning...
	Environmental technology... safety technology...
	Transport technology
	Military science & technology
	* Alternative only.
	Minerals extraction technology... oil & gas well technology...
	Process industrial technology... chemical technology...
	Manufacture & technology of special products
	* Not classed elsewhere.
	Agriculture & animal exploitation
	* Alternative only.
	Useful arts, personal services & technologies
	Household management... catering... hotels...
	Recreative arts, leisure arts
W	Arts, fine arts
W8	Styles, schools, subjects & genres
WC	Architecture as an art
WE	Plastic arts: sculpture, glyptics, ceramics...
WJ	Graphic arts: painting... reprographic arts...
WP	Decorative arts
WV	Music
WY	Performing arts
X	Philology: language & literature
XA	Linguistics
XL	Literature (general & comparative)
	Individual languages & their literature
	* As Schedule 3 (with modifications), e.g. YV French.
Y	Favoured language
	* For example, English in British libraries.
Z	Religion... occult... morals & ethics
	* Alternative is at P.

Volumes in the series

The second edition of the classification will be complete in twenty-two volumes. Thirteen have been published so far and are identified by their dates of publication.

Introduction and Auxiliary schedules. 1977.

Class 2/9: Generalia, Phenomena, Knowledge, Information science and technology.

Class A/AL: Philosophy and Logic. 1991.

Class AM/AX: Mathematics, Statistics and Probability. 1993.

Class AY/B: General science and Physics. 1999

Class C: Chemistry.

Class D: Astronomy and Earth sciences.

Class E/GQ: Biological sciences.

Class GR/GZ: Applied: Agriculture and Ecology.

Class H: Anthropology, Human biology, Health sciences. 1980. Includes medicine.

Class I: Psychology and Psychiatry. 1978.

Class J: Education. 1990.

Class K: Society. 1984. Includes social science, sociology, social anthropology, customs, folklore and mythology.

Class L/O: History. Includes area studies, travel and topography, and biography.

Class P: Religion, the Occult, Morals and ethics. 1977.

Class Q: Social welfare. 1994. Includes criminology.

Class R: Politics and Public administration. 1996.

Class S: Law. 1996.

Class T: Economics, Management of economic enterprises. 1987.

Class U/V: Technology and Useful arts (including household management and services).

Class W: Recreation and Arts. Includes music.

Class X/Y: Language and Literature.

The Bliss Classification Association

Registered charity no. 270580

The Bliss Classification Association is an association of users and supporters of the Bibliographic Classification. It promotes the development and use of the classification, publishes official amendments, enables users to keep in touch and exchange experience, and gives them a say in the future of the scheme. It is a nonprofit organisation, founded in 1967, with members all over the world.

An Annual General Meeting is held each November in London. The Association holds occasional other meetings, organises training in the use of the classification, raises funds, and publishes the Bliss Classification Bulletin each winter.

The Bulletin carries amendments to BC2, news about the Association and the development of the scheme, articles on classification, and reviews. It is available only to members. Membership is open to all organisations and individuals interested in BC2.

The Association's address is:

Bliss Classification Association
The Library
Fitzwilliam College
Huntingdon Road
Cambridge
CB3 0DG
United Kingdom

The Association is pleased to hear from users and supporters of the classification and to answer questions about it. Please address enquiries about membership or subscriptions to the Hon. Treasurer, and other enquiries to the Hon. Secretary at the addresses given below.

****

Mrs Heather Lane
BCA Secretary
The Library
Sidney Sussex College
Cambridge
CB2 3HU

hel20@cam.ac.uk

Ms Angela Haselton
BCA Treasurer
Tavistock Library
Tavistock Centre
120 Belsize Lane
London NW3 5BA

haselton@tavilib.demon.co.uk

Officers of the Association and members of the Committee

During the preparation of the present volume the following members served on the Committee elected at each Annual General Meeting to carry out the work of the Association.

Chair & Hon. Editor	
Jack Mills	
Hon Secretary	
Heather Lane	Sidney Sussex College, Cambridge
Hon. Treasurer	
Angela Haselton	Tavistock Library, Tavistock Centre
Hon. Editor, BC Bulletin	
Anthony Curwen	University of Wales Aberystwyth Dept of Information and Library Studies
Committee members	
Vanda Broughton	University College London
Sally Chambers	Senate House, University of London
Frank Emmott	Barnardo's
Marion MacLeod	Fitzwilliam College, Cambridge
Ia McIlwaine	University College London
Elizabeth Russell	Personal member
Alan Thomas	Pratt Institute, New York

Bliss Classification Association

(Registered charity no. 270580)

Subscription Form for New Members[1]

Subscription rates:	Organisations	£15.00
	Schools and individuals	£10.00

The annual subscription includes the Bliss Classification Bulletin.

Name / Job title ....

Organisation ....

Address ....

....

....

Make cheques, postal orders or international money orders payable to Bliss Classification Association. Overseas subscribers, please send international money orders in pounds sterling.

Signature .... Date ....

Return this form with your subscription to:

Bliss Classification Association
Tavistock Joint Library
120 Belsize Lane
London
NW3 5BA
United Kingdom

[1]Please photocopy this form to prevent damage to the volume

Introduction to Volume AY/B
General science and Physics

1 This volume of BC2 contains the first two classes of the natural sciences — Class AY General science and B Physics. The full classification of the sciences in BC2 will occupy seven volumes. This volume is therefore somewhat unusual in that it does not contain the schedules for the complete class (Science and the natural sciences). It contains only those for the preliminary classes in AY (mainly Processes and properties, Operations, Agents and Common Subdivisions) and for just one of the individual sciences - Class B Physics. In order to avoid duplication in the explanation of features common to science as a whole, to its preliminary class (called general science here) and to the first of its particular sciences (physics) the following order of explanation is given:

1.1 A brief statement of the purpose of this classifcation of science and of its central features;

1.2 An introduction to Class AY/I (the sciences as a whole);

1.3 An introduction to Class AY, called General science since it contains those categories of concepts common to all the individual sciences;

1.4 An introduction to Class B Physics.

2 The classification is designed to perform two major functions:

2.1 To serve as a library classification. This makes possible the organization of a comprehensive collection of documents on science in a clear and logical sequence on library shelves, in other physical media, or of entries for them in catalogues and bibliographies or other linear displays. The fundamental purpose of its logical structure is to make the locating of any particular subject within the sciences, however complex, highly predictable - the central requirement of any instrument for information retrieval.

2.2 To serve as a highly structured and detailed map of the concepts in the field of science. This may be used as the basis for thesauri or other aids for searching an information store, through the control of synonyms and the comprehensive display of the connections of all kinds between the different concepts. (Ref.1: Aitchison, Jean. Bliss and the thesaurus: the Bibliographic Classification of H.E.Bliss as a source of thesaurus terms and structure. Journal of Documentation, v.42, no.3. September 1986. pp. 160-181). This function is considered further in the Introduction to AY/I (section 10).

3 The central feature of a library classification, as distinct from any other kind of knowledge organization, is that it presents all its concepts, simple or complex, in a single, one-dimensional sequence. Most of the subject classes represented by books, journal articles, research reports, etc. are compounds of several different concepts and could equally well be classified in a number of quite different ways. For example, a work on radiography in the diagnosis of stomach cancer could go equally well under radiography, diagnostic techniques, stomach, or cancer. Such a compound reflects twenty-four different locations in a linear sequence, all of them are logically justifiable and sensible (twenty-four is the factorial product of four and represents the different combination orders in which the four concepts could be taken).

3.1 Whilst catalogues, bibliographies and other media can multiply the entries representing a document (so that the work above might get four separate entries, say, under Stomach, Cancer, Diagnosis and Radiography) this option is not open for the physical arrangement of the documents: a document can go in one place only. If the library user is to locate any given subject class easily, the sequence of classes throughout must reflect strict rules as to the combination order in such classes. So the work above might be located as Stomach - Cancer of - Diagnosis of - using Radiography, and nowhere else. Applying such rules minimizes the central (but uavoidable) weakness of the linear order, which is the scattering of some concepts by subordination to others (as works on cancer will be scattered under the different parts of the body in the order suggested above).

3.2 Whilst the development of such rules is an absolute necessity if the location of any given class is to be highly predictabe, it is also the great strength of library classification as an instrument in retrieval. The harsh demands made by the need for comprehensive rules for a consistent order of combination force the designer of the classification to examine closely the categories into which the many concepts fall and the multifarious relationships between them more rigorously than is the case with any other system of retrieval. The exposure of synonymity or near-synonymity between terms is only one by-product of the rigorous mapping of the semantic (generic and partitive hierarchical) and syntactic (non- generic) relationships between them. It is no exaggeration to say that the modern faceted classification is the most comprehensive and sophisticated organization of knowledge to be met with in the field of information retrieval.

4 BC2 schedules are the result of a rigorous and detailed analysis of the vocabulary of each of the subjects in the field of knowledge, using the techniques of facet analysis. As such, they represents a radical revision and expansion of the first edition of the Bibliographic Classification of H.E. Bliss (BC1) (ref.2: Bliss, Henry Evelyn. A Bibliographic classification. New York, The H.W. Wilson Company, 1940-1953.

4v.). The general reasons for making the revision so radical a one are given in the Introduction to BC2 (ref. 3: Bliss Bibliographic Classification. 2nd ed. [by] J. Mills and Vanda Broughton. Introduction and auxiliary schedules (Butterworths, 1977). The particular changes in AY/B are considered briefly in section 15 below.

4.1 The summary outline of AY/I on page 1 (after the preliminary pages) is designed to give a clear view of the basic structure of the schedules for science and the natural sciences. If it is remembered that the outline schedule is, like all BC2 schedules, an inverted one (see the Introduction to AY/I (section 5.7) the outline will be seen to show not only the general sequence of categories and their classes but also the basic operational rule in applying the classificationl This is the rule that compound classes (those reflecting the intersection of two or more simpler classes) are located under the class appearing later (lower down) in the schedule. For example, in Class AY, the subclasss Spectrographic instruments (AY7 M4) is located under Spectroscopy (AY7 M) and not under Scientific instruents (AY4; Geophysics (DGB) is located under Earth sciences (DG) and not under Physics (B); Biochemistry goes under Biology (E) and not uner Chemistry (C).

4.2 The basic operational rule demonstrated above is supported by a number of other basic rules governing both the design and the application of BC2 schedules. These rules are described fully in the Introduction to BC2 (Ref.3) and are repeated briefly in each separately published class, with examples demonstrating their application to the class in question. In this volume, these rules are briefly described in the introduction to Class AY/I Science and the natural sciences.

5 Acknowledgments

We are deeply indebted to Eric Coates, former Editor of the British Technology Index, and Editor of the Broad System of Ordering for his absolutely invaluable and unstinting help throughout the production of these particularly demanding two classes. Not only has his unique experience in the field of indexing technology been indispensable in the Operations and Agents facets of AY, but the whole structure and vocabulary of Class B has benefited from his unerring grasp of classificatory relations. His work on the schedules of the Broad System of Ordering (BSO) has been a constant reference point when wrestling with the complex relations manifested in the physics schedule. His contribution to this volume of BC2 has been incalculable.

We are pleased also to acknowledge the valuable contribution made by friends and colleagues in the Classification Research Group. CRG discussions have been a constant help and stimulus in designing the schedules. In particular, we are very pleased to thank Magda Whitrow, Editor of the Isis Cumulative Bibliography [of the history of science] for her valuable help in deciding the structure of the big history of science class in AY; Jean Aitchison, whose work on faceted thesauri (particularly her Thesaurofact for the English Electric Company) has been very helpful; Chris Preddle for his excellent section on the use of BC2 in automated systems in his revised edition of

Class Q. In the accumulation and monitoring of vocabulary we have been helped by the full schedules of the UDC and by the Unified classification and its accompanying Thesaurus of INSPEC (Institution of Electrical Engineers, 1973/1983).

Introduction to Class AY/I
Science and the natural sciences

1 Scope of the class and its place in BC2

1.1 The natural sciences study the phenomema of the natural world, using as their fundamental mode of enquiry what is generally known as the scientific method. Although philosophers of science dispute certain of its features, its essence is that close observation of the phenomena, combined with the use of induction and deduction, leads to the advancing of an hypothesis as a provisional explanation of what is observed. This hypothesis is then analyzed and tested objectively by a wide variety of procedures in which measurement of the data involved and controlled experiments, as far as these are possible, are major features.

From consideration of the evidence thus established the hypothesis is accepted, modified or rejected. Hypotheses which are thus validated may then form the basis of further hypotheses, thus building up the vast corpus of what is called scientific knowledge. Those parts of it which have been validated beyond any reasonable doubt take on the status of scientific laws. But it is fundamental to the idea of scientific enquiry that if further evidence demands the modification or even abandonment of a theory, however immutable it may seem to be, so be it. This may be a question of limiting its applicability rather than falsifying it, as in the well-known example of Newton's laws of gravitation being modified by the theory of relativity. But whatever the nature of the revision, the readiness in science to accept criticism and revise opinion makes science the most modest of the major fields of knowledge in the claims it makes for its product.

1.2 When the classification of the vast and endlessly multiplying body of scientific knowledge is considered, the principle of gradation, supported by and reinforcing the concept of integrative levels, has proved to be a powerful and effective instrument. It produces a major organization of the field whereby phenomena are grouped into particular sciences which can then be presented in a sequence in which each successive science builds on the preceding ones. Successive sciences then study phenomena at higher and higher levels of complexity, in the sense that they require for their explanation not only knowledge of preceding classes but also of new or emergent phenomena. Bliss called this ordering 'gradation in speciality'. Thus, energy and matter at its most fundamental level is the subject matter of physics. Organized into more complex forms, at the molecular level, it gives the subject of chemistry and this is therefore regarded as being more 'special' than physics in its scope. At more complex levels, molecular aggregations give celestial bodies and planetary systems (the subject of astronomy and the earth sciences) and these constitute environments for the development of yet more complex forms in living matter.

1.21 Human biology, the science of the most complex animal in the evolutionary progression, at present seems to mark the end of the line for the indisputably natural sciences. Whilst the demonstrably physical phenomena studied in anatomy and physiology are still amenable to the full rigour of the scientific method, the study of human behaviour, the central phenomenon underlying all the social sciences and human studies comprising Classes J/Z in BC2, has proved far less so. Whilst the principles of rational enquiry, based on logic, are still fundamental, the complexity of the relationships evident in human behaviour have not as yet allowed anything like the predictability which is taken as axiomatic in scientific enquiry (see Ziman, John. Reliable knowledge; an exploration of the grounds for belief in science. Cambridge University Press. 1977).

1.22 Although the natural sciences display systems of great complexity (eg the near-chaos found in the behaviour of our planetary atmosphere) these are all reducible to systems of atomistic facts, reflecting energy forms, particles, atoms molecules, molecular aggregates, minerals, planetary bodies, etc. The nature and behaviour of these constituents allows accurate measurement and the development of models (explanatory theories) which can be tested, disproved or validated to form a highly reliable body of knowledge

This is not the case with human behaviour and the great systems of artefacts (from social organizations to semiconductors) and mentefacts (both intellectual and imaginative) which are the products of human behaviour. No equivalents of the atomistic and eminently quantifiable phenomena found in the natural sciences have as yet been isolated in the human studies (as Bliss called them). Although psychological and sociological theoretical constructs abound and some of the social sciences (especially some parts of economics) allow some degree of quantification, there is nothing remotely approaching the predictability which marks those throughout the natural sciences.

However, it should be remembered that on the scale of human history these studies are still in their early infancy, as is science itself, however lusty that infancy may seem already.

1.23 Another classificatory feature which marks the natural sciences but is much less evident in the social sciences and humanites is that of gradation. Bliss argued that sociology (interpreted as the basic theorizing science on which all the succeeding social sciences draw) was analogous to biology in its relation to the biological sciences (in BC2 parlance, the Properties and processes facets in relation to the Part and

Types). Apart from this, no claim was made in respect of gradation amongst the other classes of J/Z.

It should be remembered, however, that a principle like gradation has been used in BC2 not as an assertion of a particular interpretation of the nature of the sciences but as an instrument of proven value in the organization of knowledge. It contributes significantly to an organization which is readily communicable, operates with exceptional consistency, and facilitates the locating and relating of classes of knowledge with a high degree of predictability - a major virtue in an instrument for information retrieval.

1.3 In the metascientific fields of knowlege which Bliss characterized as the human studies, the subject of technology (covering most of the physical artefacts of humanity) has a particularly close relationship to science. This is reflected in the fact that the general science class (AY2/AYY) is preceded (at AY1) by an even broader class, science and technology in general. The appearance of this class at AY1 is entirely a reflection of literary warrant and not of the principles of gradation or integrative levels. There is a symbiotic relationship between science and technology. It is well know that many significant developments in science have waited on the invention of artefacts to assist the process of scientific enquiry. Major examples are the dependence of optics on the development of effective optical instruments and, in turn, the dependence of other sciences on these (of biology on the microscope, of astronomy on the telescope and so on). The reverse process, in which scientific discoveries prove the basis for the development of new technologies, is equally well known and has been particularly conspicuous in the past two centuries.

1.31 This close relationship does not mean that technology should be regarded as applied science. Much of it is not particularly 'scientific' and its objectives are quite different. The cental objective of science is to elucidate the nature of phenomena dispassionately and objectively. The purpose of technology is to serve the material interests of humans. The principle of gradation implies not only that a clear distiction be drawn between technology and science but that the latter, as a product of human social activity, should file after those studies which have the structure of society and its major social processes as their subject

1.32 Nevertheless, having acknowledged these differences, the fact remains that some of the products of technology play an important and even crucial role in science. It is not merely that scientific instruments (a major part of the Agents facet in general science) are primarily the product of technology. There is also the fact that some of them raise acutely the distinction between natural phenomena (the legitimate object of scientific enquiry, demanding a location within science itself) and applications of these (which are not). Whilst some artificial products are readily accepted as quasi-natural phenomena (eg the transuranic elements, which are artificial radionuclides, produced by bombarding heavy atoms with high-energy particles) this is partly because their production is primarily an element in the scientific study of the closely related natural phenomena. But many occasions now arise where natural phenomena are subjected to highly sophisticated and contrived manipulations designed primarily to serve utilitarian human purposes; an obvious example concerns the application of optical phenomena associated with holography. Whether such concepts should be regarded as part of physics or of physics-based technology poses a problem in the classification of science and technology. This problem is considered further in the introduction to Class B Physics (section 10).

1.33 The conclusion reached by Bliss after considering such problems was that technology should be a separate class, following the social sciences and collocated with economics, the social science concerned with the production and distribution of society's material wealth. However, Bliss provided alternative locations whereby a library could collocate a technology with its dominant scientific base in cases where a close relationship held. BC2 continues this policy but with an important modification. In the case of biological technology, the study of the science and its possible applications is so intimately connected that a complete separation would be both extremely difficult and unlhelpful. This is particularly true of medicine; here, problems of human biology which are indisputably 'natural' phenomena and the legitimate province of the science are nevertheless studied primarily because of the utility that knowledge of them would have for human welfare. A classification of medicine which did not embrace human biology would be a very defective one. So the biology classes in BC2 (E/I) include substantial portions of applied science - ie technology.

2 Structure of Class AY/I

2.1 All classes in BC2 are designed consistently according to a basic pattern which reflects the six fundamental features of a modern documentary classification. In the design operation, these six features are taken (analogously to the principle of gradation) in an invariant order in which each step depends on the proeceding ones, but not vice-versa. The steps are, in order:

2.11 Organizing the terms into broad facets;

2.12 Organizing the terms in each facet into specific arrays (sub-facets);

2.13 Deciding the citation order between the facets and between the arrays;

2.14 Deciding the filing order of the facets and of the arrays within them;

2.15 Adding a notation, in which every class is represented by a symbol possessing ordinal value, to faciitate locating the class in the file;

2.16 Adding an alphabetical index, whereby a user can go directly from the name of a class to its position in the notation and file.

2.2 The theory underlying these feature is explained in detail in the Introduction to BC2 (Ref.3). Here, the structure of Class AY/I is described in the same order of fundamental features and it is assumed that users of this class will familiarize themselves with the essentials of the theory explained in the Introduction.

3 Facet structure of Class AY/I Science and the natural sciences

3.1 The main feature of the schedule is a strict adherence to the principles of facet analysis. A facet consists of the sum of classes produced when the vocabulary of a subject is divided by one broad principle of division. So the terms making up the vocabulary of science are initially organized into ('divided into') broad facets, so that terms representing concepts which all stand in the same broad relationship to the containing class are found in the same facet. For example, all terms reflecting the notion of a natural phenomenon investigated in science (eg energy, matter, molecule, star, planet, atmosphere, river, plant, animal) are brought together in an Objects of scientific enquiry facet. All terms reflecting a subsystem of any of the above (eg the lithosphere, hydrosphere and atmosphere of a planet, the regional parts or organs of an organism) are brought together in a Subsystems (Parts) facet - and so on.

3.2 Facets in Class AY/I

The facets identified are summarized below; their scope and relations are considered in more detail under citation order (section 4.5).

3.21 The entities and systems defining the objects of scientific enquiry; eg energy and matter, molecular systems, stars, planets, living things.

3.22 Subsystems and parts of the entities; eg lithosphere, nervous system.

3.23 Properties and processes of entities and subsystems; eg distribution, dimension, deformation, diastrophism, respiration, reproduction.

3.24 Operations: actions performed on all the above by human agents - eg recording, measuring, visualizing, analyzing. Operation are distinct from processes, which are activities internal to a system, as in the examples in 3.23.

3.25 Agents of processes and of operations; eg catalysts, equipment, instruments, materials. This facet also includes scientific personnel.

3.26 Common subdivisions (CSD); concepts which are to be found in all subjects and which refer largely to the human study and practice of the subject and the conditions (eg of time and place) under which is or has been pursued.

3.3 Arrays within facets

3.31 Most facets contain terms which reflect different specific principles of division, whereas a facet as a whole reflects only one broad principle of division. For example, in biology, a major facet is that of Types of organisms; these may be further divided by specific principles like taxonomic status, habitat, sex, age and so on.

3.32 This operates at every level of the classification; eg as specific a concept as fluid flow (itself reflecting the intersection of an entity, the state of matter Fluid, and a process, Flow) may still be divided by a large number of different specific principles of division, giving such classes as conical flow, compressible flow, viscous flow, laminar flow, etc.

3.33 Terms in an array are mutually exclusive, so that there is no question of compoundsing between them; eg there is no class of turbulent laminar flow. So the crucial problem of citation order between the components of a compound (see section 7) no longer arises within an array - only between them (eg to give a compound like viscous laminar flow).

4 Citation order (combination order)

4.1 This refers to the order in which the elements of a compound class (one consisting of more than one element, whether from different facets or from different arrays) are combined (cited) in a heading; eg whether the heading (which reflects the order in which the classes and subclasses are taken) is

Animals - Birds - Aquatic birds - Charadriiformes - Behaviour
or
Animals - Birds - Behaviour - Aquatic birds - Charadriiformes
or
Animals - Behaviour - Birds - Aquatic birds - Charadriiformes
or any of the other 21 permutations possible here.

4.2 Citation order reflects the order of application of principles of division and determines which concepts are subordinated to which; eg using the first heading above would scatter literature on animal behaviour according to the various types of animals.

4.3 If a consistent citation order is followed, the scattering of some subjects because of their subordination to another (an inevitable feature of classification applied to a linear order, whether of documents on library shelves or entries in manual catalogues) is strictly controlled and the location of quite complex classes (reflecting the compounding of several different facets or arrays) is always predictable. The retrieval of the information on the scattered classes is thus ensured.

4.4 Citation order is the most important single feature of a classification system. But clear and consistent rules for it can only be expressed in terms of the facets and arrays involved - hence the prior need to organize terms into their facets and arrays (see the order of operational steps in section 2.1).

4.5 Citation order between facets

In all its classes, BC2 seeks to observe as far as possible the 'standard' citation order. In any subject, this takes as the primary facet (the first-cited one) that facet which reflects the ultimate purpose or object of study in the subject, and within which the other concepts and their facets are defined. Each class in the facet may then be divided into the following facets and in the following order:

Types, Parts, Properties, Processes, Operations (actions on it), Agents (of processes and operations), Common subdivisions

The following notes explain how these general rules have been applied to Class AY/I.

4.51 The primary facet consists of the entities defining the objects of scientific enquiry. They embrace fundamental manifestations of energy and matter (kinetic energy, forms of motion, waves, electromagnetism, particles and more complex aggregations of these (molecular, astronomical and planetary, living organisms, etc). The concepts making up these classes are easily recognized as defining the well-known sub-disciplines of science (physics, chemistry, etc).

4.52 The second-cited facet consists of the Subsystems and Parts of these; this facet has no general class independently of the entities above, only those subsystems special to a given entity and which vary widely according to the entity. It includes subsystems of chemical molecular systems, parts of planets such as lithospheres and atmospheres, the regional parts and organs of living things and so on.

4.53 The third-cited facet consists of Properties and Processes of systems and subsystems. These range from ubiquitous concepts appearing in all systems (eg distribution, variation, frequency, dimension) to specialized ones such as oxidation and reduction in chemical substances, tectonic and degrading processes in planets, and respiration, evolution, etc. in living things.

4.531 Properties are conceptually distinct from processes (actions internal to a system); eg colour is a property, change is a process.

4.532 Where the distinction is clear, two different facets will be recognized; but in many fields the distinction is so blurred that the two facets are conflated. In terms of citation order, property is always subordinated to whatever it qualifies, whether this is an entity, a process or even another property (eg the durability of a colour).

4.54 The fourth-cited facet consists of Operations; these are actions performed on all the above by human operators and their agents, as distinguished from processes, which are activities internala to a system, such as the physiology of living things.

4.55 The fifth-cited facet consists of Agents; these are usually agents of external human Operations, such as equipment and instrumentation. But the relationship of action/agent also occurs with processes (eg enzymes acting as biochemical agents in a physiological process).

4.56 The sixth-cited facet consists of Common Subivisions (CSD); these are concepts which are to be found in all subjects and provided for in BC2 by Common Auxiliary Schedules 1/4, applicable to all classes. They range from bibliographical forms (dictionaries, graphic materials, etc) to operations like study and research and agents of these such as organizations.They also include the two major facets of Time and Space in their commonly occurring manifestations of chronological periods and geographic places. The concepts of Time and Space as fundamental parameters in natural phenomena are regarded as subject classes in physics (at B9B).

4.6 Citation order within facets (between arrays)

4.61 In contrast to citation order between facets, there are no general principles available for deciding citation order between arrays. Decisions are largely empirical, based on considerations of whether any given compound (reflecting two or more different arrays) would most helpfully go.

4.62 The number of different arrays is often so large that it is out of the question to list them in citation order as is done for facets in section 5. However, the order in which they should be cited is shown clearly by the inverted filing order (see section 5.2 below); an array filing later (further down) in the schedule should be cited before one filing earlier (see examples in section 5.22).

4.7 General indexing rules for citation order

4.71 The rules described above govern by far the greater number of decisions for compounding in BC2. However, a number of well-established indexing rules, all of them consistent with the standar citation order, are also observed and are very useful in practical classification (see Introduction to BC2 (Ref. 3) section 7.331). Sometimes, these demand that synthesis should be by building forward, not retroactively. This is because the normal relationship between the facets or arrays has changed. The most prominent of these is probably the rule for citing in the order Patient (ie recipient of action) - Action - Agent. This is usually taken care of by the normal citation order; eg Techniques in scientific investigation AY6/AY7 file after Equipment and Materials AY3/AY5 and are cited before them. But when one thing influences another (a special case of the agent relationship) the influencing factor, which is cited second, may file after the thing affected; eg thermal acoustics BRG HGP is built by citing acoustics (BRG H) before thermal properties (BRG P) although acoustics (here, the thing affected) files before thermal properties.

4.72 The situation may be generalized thus: whenever the relationship between concepts varies from that embodied in the standard citation order, these general indexing rules should be invoked.

5 Filing order

5.1 This is the order in which the individual classes, simple or compound, file one after the other, whether in the schedule, on the shelves or in a catalogue or bibliography. It has two quite separate components - facet filing order and order in array.

5.2 **Facet filing order**

5.21 This is the order in which the different facets, each one containing a block of different classes, file one after the other.

5.22 All schedules in BC2 are inverted ones; ie the facets file in an order which is the reverse of the order in which they are cited when compounding terms to form compound classes. So the primary facet files last, the second-cited facet files next to last, and so on.

5.23 The reason for this (see the Introduction to BC2 (Ref.3) is solely to preserve a consistent order of general-before-special. The assumption that a general class should file before its subclasses is virtually universal.

5.24 Example of an inverted schedule:

G	Zoology, animals
	(Processes)
GHT	Behaviour
	(Types of animals)
	(Types by development)
GL	Young
	(Types by habitat)
GN	Aquatic
	(Types by taxonomy)
GP	Birds
GPH T	Behaviour
GPN	Aquatic birds
GPP T	Eagles
GPP THT	Behaviour

5.25 In the file above, the compund Bird behaviour (GPH T) files after both the more general classes to which it belongs (Animal behaviour GHT and Birds GP). If the schedule were not inverted, and the Processes facet filed after the Types of animals facet (just as it is cited after it) the general class Animal behaviour would file after its subclass Bird behaviour.

5.26 Similarly, within each each facet the arrays are inverted; the first-cited array files last, the second-cited array files next to last, and so on.

5.27 It was noted in section 4.62 that the inverted filing order embodies within itself a comprehensive guide to the citation order. The sequence of classes in 5.24 demonstrates this. It implies, inter alia, that a process is cited after a type of animal; also, that within the Types facet an animal characterized by its taxonomic class is cited before its habitat.

5.3 Order in array

5.31 The classes in an array are mutually exclusive and cannot normally be compounded (see section 3.33). So their filing order cannot be determined by citation order. Where there is an obviously helpful order, that is used; eg chronological order of periods in AY7 History of science; evolutionary order, reflecting development over time, of organisms in biology; geographical order of places in DU Geography.

5.32 However, order in array is usually pragmatic. The filing order of the individual sciences is a somewhat special case and is probably best regarded as that of a quasi-facet order. The order is essentially one of gradation and integrative levels (see sec 1.3) and provides both a filing order and a citation order. Compounding between the special sciences (which is not uncommon) conforms to the basic retroactive rule, the class filing later being cited before that filing earlier; so the physics of chemical substances, of astronomical bodies, of terrestrial processes, of biological processes, etc all go under the class filing later, not under physics.

6 Alternative treatments and arrangements in the order of classes

6.1 These serve the demands of particular types of libraries. In each case, the notation has been designed specifically to allow alterations to be made to the preferred arrangement. In all cases, the preferred arrangement is stated clearly and any special notational instructions needed to implement the alternative are indented under the note for it. The general pros and cons of alternatives are explained in the Introduction to BC2 (Ref.3).

6.2 The main alternatives in Class AY/I are noted in the separate introductions to the classes concerned. One common to all the sciences is that for the collocation with its science of the applied science or technology of that science. But note that in the case of the biological sciences, the collocation of the technology with biology is the preferred arrangement.

7 Notation

This is explained in detail in the Introduction to BC2 (Ref.3), Only its main features are described here.

7.1 Notation is a system of classmarks representing the terms (classes) of a classification. Its function is to locate in a mechanical fashion the position of each and every class, simple or compound, in the system. It does this mechanically because the symbols (in BC2, capital letters and numerals) have an ordinal (positional) value already well known to the users. The only special ordinal value the user of BC2 must remember is that a number files before a letter; eg AY9 files before AYA.

7.2 The notation is purely ordinal - ie it makes no attempt to express hierarchical relations by adding another character to symbolize each step of division. Such an attempt must always fail sooner or later and more likely sooner. So BC2 notation concentrates on the primary function of notation, which is simply to maintain the order of classes already determined completely by the theoretical rules governing order (citation order and filing order). By doing this, it secures classmarks which are as brief and as simple as possible. For example:

AY3 6	Practical scientific work
AY3 B	Equipment & materials
AY3 U	Equipment & plant
AY4	Scientific instruments
AY4 5	Components
AY4 K	Switching devices

7.3 Only one classmark in the above chain (sequence of successively subordinate classes) is 'hierarchical' in that it adds a character to the classmark of its immediate containing class; ie AY4 5 can be 'seen' to be subordinate to AY4. On the other hand, the classmarks are much shorter than they would have been otherwise; eg the last class would need a classmark nine characters long if the notation were hierarchical.

7.4 It should be emphasized that notation in no way determines the order of the classification. The order is determined entirely by the theoretical principles and rules described in sections 3/6 above. All that notation does is to maintain this order mechanically; it is the servant of the order, not the master. Nevertheless, notation is an important feature of a classification; as Bliss said, it may not make the classification, but it may mar it. BC2 seeks to keep notation as brief as possible (whilst still being as specific as possible in pin- pointing classes) since brevity is the major element in simplicity.

7.5 The notation is fully faceted and synthetic. Compound classes formed by the intersection (coordination) of two or more separate concepts or classes are given classmarks which are built ('synthesized') from the simpler constituent classes according to strict rules. These are explained fully in the Introduction to BC2 (Ref.3) but the essential ones are repeated here for convenience.

7.51 The main function of synthesis in notation is to provide maximum hospitality. Within any subject, the number of potential classes is enormous. Every term in every facet is theoretically capable of intersecting with every term in every other facet. Although the literature may reflect a large number of these, this number is still only a fraction of the number it might conceivably deal with. The notation must be flexible enough to accommodate all of them.

7.511 It must also, of course, be able to accommodate new concepts as they arise. It is assumed that new concepts will always fit into existing facets, which reflect categories of concepts fundamental enough to ensure this. The provision of correctly located classmarks for newly inserted concepts is also facilitated by the ordinal notation, which suffers far less from rigidity (as Ranganathan called it) than so-called hierarchical notation.

7.52 Enumeration of compound classes in the schedule

7.521 Because of the practical impossibility of printing out ('enumerating') in the schedules all the compound classes which may arise, a faceted classification may decide to adopt a rigorous policy of not giving any. This was the case in the first fully faceted classification to be made - the Colon classification of S.R Ranganathan. This simply listed all the elementary terms in their facets and left it to the classifier to build the classmarks for all compound classes as they appeared in the documents being classified.

7.522 For a number of reasons this schedule style is not followed in BC2, which enumerates a fair number of compound classes. The main reason is that this assists the classifier to grasp the structure of the whole class as it affects the particular class concerned. It shows clearly how the the concepts in a subject are handled consistently despite the ambiguities and confusions in the terminology. These are far more prevalent in the case of the natural sciences than is warranted by the description of their terminology as being 'hard' (as compared to the 'soft' terminology of the social sciences). Wherever necessary, definitions of terms are given so that the classifier can see why the concept has been classified as it has been. Enumeration of compound classes also facilitates the provision of A/Z index entries for important compound classes (which would not get an entry were the schedules to be confined to the elementary terms).

7.523 In such cases (of some enumerated compound classes appearing in the schedule) it should not be thought that the detail under that part is limited to the subclasses thus enumerated. When assessing the specificity of the vocabulary in a given class it should always be remembered that the class may be qualified by all earlier **facets and** arrays, whether this is hinted at by a limited enumeration or not. For example, there is enumeration of the class Neutron-neutron interaction at BNW QW; but this could easily be notated if the need arose.

7.6 Classmark building (synthesis)

This is best demonstrated by detailed examples from a specific class; so the explanations in the introduction to Class B Physics may serve to demonstrate the problems for all the sciences. Here, only the briefest summary of the position is given.

7.61 The chief method of synthesis is by retroactive notation. It must be remembered that compounding in BC2 is nearly always done by qualifying a given class by one or more other classes preceding it in the schedules. The classmark for the compound is obtained by adding the earlier classmark directly to the later one after dropping any initial letters (or, much less commonly, numbers) common to the two classmarks being joined; eg

BF Waves

BM	Particles
BMF	Particle waves

Here, the earlier class (waves) is added directly to the later one (particles) after first dropping the initial B, which is common to both of them.

7.611 A distinction worth noting here is that of a notational facet, as distinct from the more commonly referred to conceptual facet. A notational facet is the set of subclasses beginning with the same initial notation; eg the notational facet AY7 consists of all the subclasses beginning with AY7 (whereas the conceptual facet of Techniques in scientific investigation in which it appears covers AY6/AY7). This distinction is useful when considering how many letters can be dropped in retroactive notation.

7.62 The second method of synthesis uses an intercalator. This is any letter or number used to introduce subclasses other than by the automatic operation of retroactive notation. It is always accompanied by an Add instruction; eg

AY3 B	Equipment & materials (in scientific investigations)
AY3 K	(Physical properties of equipment & materials)
	* Add to AY3 K letters A/W following B; eg
	AY3 KGP Thermal properties [from BGP]

7.621 An important use of intercalators is when every member of a large class has to be qualified by a standard set of subclasses (eg every particle in physics, any given element in chemistry).

7.622 Intercalators are quite prominent in Class B Physics and the Introduction to Class B Physics (section 8.4) gives further examples.

7.7 The different ways of building classmarks described above may give an impression of complexity at first reading. But so would the simplest action if described in terms of its basic operational steps. The detailed sequence of instruction we need to give a computer before it can process the simplest operation demonstrates this. Applying notation is a practical operation; the steps involved are basically simple and quickly become familiar after a little practice.

8 The alphabetical subject index

The function of the A/Z index to a classified indexing system is considered in the Introduction to BC2 (Ref.3) (section 6.5 gives general principles and section 7.5 gives practical guidance for a library making its own A/Z index to its own stock). Only the basic features are given here.

8.1 The main points to be remembered for the efficient use of the printed index to the schedules in this volume are given on the page preceding the A/Z index.

8.2 The A/Z index to this volume is essentially a quick guide to the location of any given concept in the schedules. It is important to remember that it is not a substitute for the categorized and hierarchical display of the schedules. The classifier should never classify solely by the index, but turn to the schedules to verify that a given context is the correct one for the concept as used.

8.3 Most of the terms indexed are elementary ones, whereas much of the literature to be classified reflects several such terms combined in some form. The classifier must be thoroughly familiar with the basic rules governing how these terms are put together in order to get a classmark which locates the compound accurately in its correct context.

8.4 The basic rules followed in making the index are those of chain indexing. This is a highly economical way of reducing the number of entries needed for compound classes whilst ensuring that all keywords likely to be sought by a user of the index will still appear. The main rule is very simple: if an entry term is qualified at all, it is by a superordinate term (ie a containing class, which helps to define it). An entry term is never qualified (followed by) a term representing one of its own subclasses in the schedule; eg entries may appear for

Propagation of waves BFC
Waves BF
but NOT for
Waves, Propagation BFC

since the last class represents a subclass of waves in the schedule. Should a user of the index consult it under the last form, and not find it, they will nevertheless find the direction to BF Waves (in general) and can locate the desired subclass in a slightly less direct way via the schedule.

8.5 The A/Z index to BC2 classes is now produced largely by automatic selection of terms from the schedules, using a computer program written to this end. This include, for example, rules for deleting 'anti-chain' entries; this ensures that no entry appears for Wave propagation (say).

8.6 A problem which will need resolving when BC2 is complete is that of a consolidated A/Z index. The obvious desirability of one A/Z index to all the sciences and not just to AY and B anticipates the more general problem. The Bliss Classification Association has begun looking at this, but until BC2 is complete the actual production of such an index is, unfortunately, not practicable.

9 Practical classification in AY/I

This considered in detail in the Introduction to BC2 (1979). Some brief notes on the problem, with demonstrations of the different situations which arise, will be found in the Introduction to Class B Physics.

10 Applications of Class AY/I

10.1 The use of BC2 to arrange files of documents or catalogue entries for them, including the provision of a conceptual framework to assist the analysis of the documents, should be clear from the various sections above. But BC2 is also extremely useful in the construction and searching of automated as well as manual databases.

10.2 Automated catalogues and databases

10.203 Reference has already been made in the Introduction to Volume AY/B (section 2.2) to Jean Aitchison's paper in the Journal of Documentation on the use of BC2 in this context. An even more detailed account, albeit one restricted to a single subject field for its examples, is given in the introduction to Chris Preddle's revised edition of Class Q Social welfare and criminology (Bowker-Saur, 1996). It considers the problems of subject keyword indexing, the precoordination of terms, indexing aids in controlling preferred terms, synonyms, etc, subject keyword searching, thesaurus construction and searching for documents by searching classmarks.

10.21 Ideally, the substance of the above would be repeated here, but using the vocabulary of science for its demonstrations. However, the exigencies of getting this first volume of the BC2 classification of the sciences published without further delay have made this impracticable. But although the examples used in the Class Q Introduction are drawn from a quite different field of knowledge, the principles involved are much the same as will be found in any field and their exposition should proved of great help to users of this volume until an exposition tailor-made to the needs of the sciences is available.

10.3 The use of BC2 in constructing traditional manual catalogues is considered in the Introduction to BC2 (Ref.3) and will not be repeated here. Excellent comments on this problem will also be found in the Introduction to Class Q.

11 General science and physics in BC2 compared with BC1

11.1 The general sequence of the sciences in BC2 is virtually the same as in BC1. This is not surprising, since the main-class order has always been recognized as the greatest strength of BC1. But within each individual science, the differences between BC1 and BC2 are very substantial indeed, both as to the organization of classes and the size of vocabulary. These differences in the individual sciences are considered under each science and only AY and B are considered here.

11.2 Class AY

In BC1 the subject of science in general is at AK, following philosophy (A/AJ) and preceding logic (AL). Bliss acknowledged that both logic and mathematics (AM/AY in BC1) were not sciences properly, being concerned not with the observation of natural phenomena per se but with general methods applicable to all subjects, and particularly to philosophy and science. Nevertheless, he described them as abstract sciences which made a bridge between philosophy and science proper. BC2 has interpreted the scope of the latter as it is more generally understood, as consisting of the empirical sciences concerned with 'objective realities' as Bliss put it. By this reckoning they are more concrete than logic and mathematics and therefore follow them in main-class order.

11.21 The small vocabulary of AK in BC1 consists mainly of relatively generalized terms which tend to mix up concepts usually regarded as beloging to the common subdivisions (eg surveys, recreations in science), social relationships of science (eg science and theology, science impostors, applications to industry) and scientific operations (eg organization of research). Only one classmark is given to the agents of scientific activity (laboratories and equipment) and this is separated from scientific instruments, which is preferred under physical sciences (AZN). There is only one concept representing the general processes and properties (AKS Symmetry and conformity in natural objects). Although AK contains several concepts relating to different categories of sciences (eg exact sciences, formal sciences, natural sciences, sciences grouped for some purpose) the general concept is not completed until the widely separated class for physical sciences at AZ. So AK is not used in BC2 and all its concepts are contained in AY.

11.22 The overall order of Class B Physics in BC1 is broadly similar to that in BC2, but lacks consistency in its categorization. This is best shown by outlining it:

BA/BB	Theoretical & practical physics
BC/BD	Mechanics
BE/BF	Matter & energy... Radiation
BG	Properties of matter, changes of state
BH/BL	Heat... Light... Electricity & magnetism...
BM/BO	Electrical techology & Engineering
BP	Sound
BQ/BR	Hydromechanics & hydraulic engineering
BS/BT	Aerodynamics & aeronautics
BU	Physical technology

11.221 In terms of facet structure this shows some confusion. Although BG, for example, includes the general classes for the particular states of matter (which are given in the same order as in BC2 - Fluids... Gases... Liquids... Solids) the content of these classes is scattered (at BQ/BT) with little apparent rationale. No clear distinction is made between the particular states of matter and the processes concerned in their changes.

11.222 The general state of this internal structure in the class is one reflecting the state of library classification before Ranganathan introduced the strict application of the rules of logical division to sort out categories of concepts and developing clear rules for their interaction.

11.223 The other striking difference between BC1 and BC2 in the two classes is the size of vocabulary. It is very difficult to give even an approximate figure for the size of vocabulary in a synthetic classification, since the compounding of classes can generate tens of thousands of classes if the literature demands it.

11.224 If we consider only those terms enumerated in the schedules with a specific classmark, BC2 has something like twenty times as many terms as BC1 in the general science class and some twelve times as many in physics. The figures (very roughly indeed) are: for general science BC1 has (in AK and AY) some 50 enumerated classes of which some 40 have individual classmarks. BC2 has (in AY) some 1100 enumerated classes. These figures exclude the Common subdivisions, or the disparity would be even greater. In physics, BC1 has some 600 enumerated classes, of which only about half have an individual classmark; BC2 has between 3500 and 4000 enumerated terms, all with individual classmarks.

11.225 When it is remembered that BC2 is a fully synthetic scheme, whereas BC1 has very limited synthesis in its two classes, these differences in range of vocabulary increase more or less beyond measure.

Introduction to Class AY
General science

1 Basic principles and structure

1.1 The underlying principles governing the classification of all subjects in BC are described briefly in the Introduction to Class AY/I (sections 2/11). They include definitions of facets and arrays and basic rules for citation order and filing order and for notation.

1.2 The Outline of Class AY (p.1) is designed to give a clear view of both the scope and the basic structure of the class. Since the schedule is an inverted one (see section 8.2) the outline will be seen to show not only the general sequence of categories and their classes but also the basic operational rule in applying the classification. This is the rule that compound classes (those reflecting the intersection of two or more simpler classes) are located under the class appearing later (lower down) in the schedule. For example, Recording is AY7 4T and Holographic techniques is AY7 K; so recording techniques in holography is AY7 K74 T.

2 Scope of Class AY and its place in BC2

2.1 The facets in the complete class AY/I (Science and the natural sciences) are summarized in the Introduction to AY/I (section 3) and explained more fully in sections 5.5 (under citations order). This shows that the main body of the individual sciences, dealing with the different types of natural phenomena studied, together with their subsystems, occupy classes B/I. Collectively, they make up the primary and secondary facets of science.

2.2 However, a small residue of concepts belonging to Facets 1 and 2, but more general in scope than any of the classes B/I, are located in AY. The full facet structure of AY can now be summarized.

3 Facets in Class AY:

3.1 Objects of scientific enquiry & their subsystems (AYE/AYR)

3.11 As noted above, nearly all the classes in these first two facet of science are accommodated in B/I and only a small residue of more general concepts is found in AY: ie

3.12 Aggregations of the individual sciences (eg empirical sciences, fundamental sciences, physical sciences);

3.13 Systems and subsystems in general: this class is on the borders between the categories of entities studied, structural properties of those entities and conceptual tools for the study of these. One feature of a class on systems and subsystems in general is that it may be equated with the idea of complex systems in general and it is this sense that makes its location immediately preceding the individual, special sciences both logical and helpful.

3.14 Although the application of systems theory is usually in the area of the natural and social sciences, it is also used in the applied social sciences, particularly in the field of management, and to technology (in systems analysis, systems engineering, etc). In this relationship, the locating of systems here is consistent with that of other concepts with a base in science but applied extensively outside it.

3.2 Properties and processes (AY9 2/AY9 R)

3.21 The contents of this facet vary considerably from one individual science to another. For example, in DG Earth sciences, they include seismic and volcanic action and processes of erosion and deposition in the lithosphere; in the hydrosphere and atmosphere quite different concepts have to be accommodated. In E Biology, processes like the evolution and development of species and the physiology of individual organisms are found.

3.22 In AY, enumeration is restricted to those general concepts which are to be found in all or many of the individual sciences. These include concepts like distribution, change, dimension (in time and in space) and others from systems theory, such as systems behaviour, constancy, reversibility, etc.

3.23 A facet containing terms of high generality which are applicable to many different contexts is likely to clash on occasions with the special provisions found in those contexts. For example, the array for Conditions & parameters (AY9 4/AY9 7) duplicates the provision for a few ubiquitous concepts, such as temperature and pressure; they are designed to be used when qualification by the general classes in the large Energy interactions & forms facet is unwieldy and the wide context implied by these terms is not required. But if there is some doubt as to which is best for a particular situation, the general class should be preferred.

3.3 Operations (AY2 YM/AY8)

3.31 These reflect the actions of humans in their scientific study of phenomena. They constitute a major facet in AY and two major quasi-arrays are distinguished: theory, representing purely intellectual operations, and practical scientific work and techniques. Strictly speaking, both these are subclasses of the one overarching operation of scientific research.

3.4 Agents (AY3 B/AY5)

This facet consists almost entirely of the equipment and instrumentation used to assist the operations of enquiry. A few agents (notably the persons themselves) will be found in the Common subdivisions.

3.5 Common subdivisions (CSD)

These are common to all subject field, not only to general science and the individual sciences. Conceptually, they constitute a special application of the relationship of operations and agent in which the operations relate to the social organization of the study and practice of the subject and the agents are persons and organizations. Its most prominent class is probably that of the history of science. where the history relates not to the phenomena which are the object of scientific study but to the study itself, as manifested over historical time and in different places.

4 Arrays in Class AY

Arrays are sets of terms within a facet which reflect a specific principle of division as distinct from the broad principle which is demonstrated by all the terms in the facet. For example, in the CSD, Persons in science (AY2 4) are divided by sex, by ethnic group, by occupational characteristics, etc. Operations in scientific investigation are divided by scale (AY6 6), by energy form utilized (AY6 9), by the functional action performed on phenomena (AY7) and so on.

5 Citation order

5.1 Citation order of facets The citation order between the facets in AY is the order in which the facets are set out in section 3 above. This reflects the standard citation order. The Objects of scientific study are part of the primary facet and are therefore cited first. The classes in the Properties & processes facet are integral features of the natural phenomena studied in science and are therefore cited next, before the Operations on them and the Agents of those operations. CSD are cited last, as in all BC2 classes.

5.2 Citation order of arrays

The distinction between citation order of facets and of arrays is explained in the Introduction to AY/I.

5.21 Although there are no comprehensive rules governing citation order in array, the practical rule of retroactivew synthesis is readily followed by citing first the class appearing later in the schedule; eg the array of Operations by energy array (AY6) files earlier than the array of Operations by purposive function (AY7). So a compound on electronic techniques in imaging would be AY7 I6I B, not AY6 IB7 I.

5.22 Where the standard citation order can be applied, order in array reflects this; eg under Persons in science (AY2 4) the array (By occupational characteristics) files last, since it reflects a functional definition, and would therefore be cited before the others.

5.3 The basic indexing rules described in the Introduction to Class AY/I (section 5.7) are likely to be invoked in AY rather more often than in other classes; eg

5.31 Common subdivisions have always been distinguished by the fact that citation order may sometimes call for forward building; eg the history of anything is normally subordinated to that thing; so a history of scientific societies would go under the societies, although these file earlier than history. But a society devoted to the history of science would go under history of science.

5.32 In the same way, some operations, like measurement and evaluation, will always be cited after the thing acted on, even if this calls for building forward rather than retroactively; eg AY7 4N7 D Telemetry - Evaluation.

5.33 Although the fact that facets and arrays in AY are introduced by numbers allows forward building to be done quite easily, it should not be done in cases where provision is already made for the special relationships which the basic indexing rules cover; eg the Agents facet at AY3 B has its own special Operations facet provided and the general Operations facet at AY6/7 should only be used for operations not provided for by this special facet.

6 Filing order in AY

This is inverted, as explained in the Introduction to AY/I (section 6). So the first cited facet (Objects of scientific enquiry) files last, and the last cited facet (CSD) files first. Similarly, within each facet, the arrays file in the order which is the reverse of their citation order.

7 Alternatives in AY

An example of the modification of the preferred BC2 order to meet special requirements occurs at AY2 8/AY2 9B. Here, libraries wishing to keep together all the material on the history and philosophy of science, including that of the individual sciences, may do this by using the intercalator for individual sciences at the end of the history class, following AY2 96E.

8 Notation and classmark building in AY

8.1 AY has an unusual feature compared with most BC2 classes in that all its subclasses have a number as their initial character. Since the great majority of BC2 classmarks consist mainly of and end in letters, qualification by another AY classmark (after dropping the initial AY which is common to them all) is readily done by simple addition, without the need to reserve earlier characters as is the normal case with pure retroactive notation. The change from letters to numerals clearly signals that synthesis has been effected; eg AY7 M4 clearly indicates that the classmark AY7 M (for Spectroscopy) has been qualified by the earlier classmark AY4 (for Instruments).

8.2 It has already been noted (in section 6.231) that this feature makes forward building of compound classmarks easier than it usually is. However, it should be done only to implement the citation order rules of BC2, otherwise it would undermine the predictable consistency which is a hallmark of BC2.

8.3 Apart from the two features above, notation is governed by the normal rules for retroactive notation: earlier classmarks are added directly wherever this has been provided for.

8.31 Since the initial AY is common to all classmarks in AY, this is always dropped before adding earlier classmarks; eg AY4 is Scientific instruments, AY2 7 is History of science, so a history of scientific instruments is AY4 27.

8.32 Within a 'notational facet' (all the classes beginning with the same numeral division of AY) the initial three characters (AY and the initial numeral) may usually be dropped; eg

AY3 B	Equipment & materials
AY3 BD	Design of equipment & materials
AY3 C	Handling of equipment & materials
AY3 CBD	Design of handling equipment & techniques
AY3PV	Intelligent equipment & materials
AY3R	Materials in general
AY3 RC	Materials handling
AY3 RPV	Intelligent materials

8.33 On those occasions when it is necessary (because of the basic citation order rules) to add a classmark occurring later in the schedule, the use of the initial numeral allows synthesis; eg

AY7 MLX	X-ray spectroscopy
AY7 MLX 7MNP	Electron probe microanalysis

Here, the electron probe acts as an aid (agent) to the primary purpose of X-ray spectroscopy, so it is cited after the latter although it files later than the concept qualified.

8.4 Intercalators are used where convenient to introduce concepts from other classes; eg

AY3 7	Hazards in practical scientific work
AY3 7MG	Dangerous substances
AY3 7N	Specific substances
	* Add to AY3 7N letters H/U following C in Class C Chemistry.

9 Special problems in Class AY

As far as is feasible, problems of definition and analysis and the practical decisions taken to meet them are noted in the schedule at the point of occurrence. Others are implicit to some extent in the provision of alternatives. But a number of theoretical problems not adequately explained in the space of a schedule note are briefly considered here.

9.1 AY4/5 Instruments & instrumentation

The distinction between components of instruments and types of instruments is not as clear as the usual distinction between types of a thing and the parts of those things. Most components function withins a more comprehensive system of interacting components; eg one which observes, detects, indicatea, records, and measures in one integrated system. Each functionally defined device may be considered separately, but as a component rather than as a type of instrument and so the main enumeration under instruments is of the components rather than the types.

9.2 AY6/7 Investigative techniques

A distinction has been drawn between those techniques which largely reflect a mode of operation, energy form, etc (eg mechanical, electronic) and those which reflect a particular action on the phenomena investigated. The latter may be defined by what the human operator does (observe, detect, record, measure, test, etc) or by a particular property or process of the phenomenon studied (eg its spectra, microscopic structure, etc). The distinction between the mode of investigation and the thing investigated gives not only reasonably distinct arrays but also allows the application of the standard citation order (which cites the thing investigated first).

9.3 AY7 E Standards

The location of this concept was decided by the fact that the notion of testing and evaluation is implicit in its definition (see note at AY7 A). Although standards are more often considered in a technological context than a purely scientific one, their origin seems to lie clearly in the general technique of testing and evaluation of material phenomena It should be noted that the the subject of standards in other fields of knowledge is a very different matter.

9.31 The importance of standards in the study of materials, equipment, etc has led to the development of a substantial social organization for their agreement and maintenance. This is also regarded as a distinct subject in the social organization of science and technology and is therefore located in AY1 (Science and technology together).

9.4 The filing order between theory and practice was originally considered to be determined by the general principle of increasing concreteness (formalized by Ranganathan) which would file theory before practice. But when developing the physics schedule it became apparent that this would give an unhelpful order. The theoretical constructs in physics are intimately bound up with the phenomena studied. In may cases they are akin to the systematic presentation of the various processes and properties of the phenomena investigated and are designed to explain their interrelations. If an hypothesis is a strong one, the theory may hardly be distinguishable from a statement of the real relationships between the phenomena. As a result of this, it was decided to locate theory immediately ahead of the Processes and properties facet, and filing after the practical techniques and agents.

9.41 It is arguable that physics is atypical in this respect, so an alternative is provided (at AY3 2) which files theory before practice.

9.5 The history of science proved to be an example of the need to adjust the standard citation order to the demands of a special subject. In the history of science the period is far more significant than place and the collocation of all works on the history in a given period is much more helpful than the collocation of all the work done in a particular place. So the citation oder was changed to one in which period is cited before place.

10 Practical classification in Class AY

The general problems relating to the choice of alternatives, deciding which concepts are needed to summarize the subject of a document, etc. are dealt with briefly in the Introduction to Class B (section 10) and are not repeated here. The examples below demonstrate some of the problems which may arise in conceptual analysis and the assignment of classmarks. The final classmark of the document is the one which appears at the end of the chain (eg AY4 SS3 V) for intelligent sensors.

10.1 Title: Intelligent sensor systems

This demonstrates the problem of overlap between functional operations (eg sensing) and instruments which are defined by their main operation. This work seems to be concerned chiefly with instrumental systems and is therefore placed under instruments. The resultant chain is:

AY4	Instruments
AY4 5	)Components(
AY4 S	)Input devices(
AY4 SS	Sensors, probes
AY4 SS3 V	Intelligent sensors

10.2 Title: Radiation protection of accelerator environments

This demonstrates the application of the general Operations & agents facets in AY to a specific context.

BM	Particle physics
BM7 T	Acceleration of particles
BM7 T37	)Hazards(
BM7 T37 K	)Radiation hazards(

BM7 T37 KER	)Protection(

AY7 T uses a set of classmarks reserved at AY7 T for operations special to a given context. The ER for protection is built by normal retroactive notation from AY3 7ER.

10.3 Study for a degree in science & technology

This demonstrates the use of a common subdivision, with an intercalator introducing classmarks from another main class (here, degree courses in education JFO R) and the use of the very first class in AY - Science & technology together (AY1).

AY1	Science & technology
AYI 26A	)Study & education(
AYI 26A R	)Higher education([from JR]
AYI 26A RFO R	Degree courses [from JFO R]

Some librarians would regard the concept of degree as being implicit in higher education and prefer the shorter classmark AYI 26A R.

10.4 Title: Nationalism and internationalism in science, 1880-1940: four studies of the Nobel population.

The publisher's catalogue entry for this work describes its subject as scientific activity since 1880 viewed as being part of the emergence of the nation state. Applying the general indexing rules mentioned in the Introduction to AY/I (section 4.7) this is analyzed as the influence of nationalist sentiment on scientific activity; as such, the thing influenced (scientific work) is cited first. The Nobel prize winners are clearly the study sample taken to investigate the problem. The subject could be analyzed as:

AY2 9R	Science & politics
AY2 9RA NS	Nationalism [from RAN S Ideologies]
AY2 9RA NS2 8N	20th century

The period indicated covers most of that dealt with. But a library holding a large ctock in this area might wish to be more precise. In this case, Auxiliary Schedule 4A could be use to obtain high precision, in which the starting date is given followed by the duration of the period studied (to the nearest decade). This would add to the classmark -28 (for historical period) the starting date -LPT for 1880 followed by the duration (60 years) to give -LPTT. This would exact a long classmark as the price of its high precision: AY2 9RA NS2 8LP TT.

Strictly speaking, a completely specific index description here would add the further step: Study sample - Scientific award winners - Nobel prize winners. However, a more practical solution would be to treat the Nobel population as a subject of enquiry in its own right and give it an added entry in the subject catalogue. Such an entry would be classed at AY2 9KH KM.

Introduction to Class B Physics

1 Basic principles and structure

1.1 The underlying principles governing the classification of all subjects in BC2 are described briefly in the Introduction to Class AY/I (sections 2/11). They include definitions of facets and arrays and basic rules for citation order, filing order and notation.

1.2 The Summary outline of Class B (p. 20?) is designed to give a clear view of the basic structure of the class. Since the schedule is an inverted one (see section 8.2 of the Introduction to AY) that outline will be seen to show not only the general sequence of categories and their classes but also the basic operational rule in applying the classification. This is the rule that compound classes (those reflecting the intersection of two or more simpler classes) are located under the class appearing later (lower down) in the schedule. For example, BAF is Energy interactions & forms; BM is Particles; so Energy interactions of particles is BMA F - ie under the later-filing Particles, not the earlier-filing Energy interactions in general.

2 Scope of Class B and its place in BC2

2.1 Broadly speaking, physics may be defined as the study of matter and energy aimed at describing natural phenomena in terms of fundamental laws. Using the terminology of gradation and integrative levels (see Introduction to AY/I, section 1.2) this means that it considers natural phenomena at their most elementary level, before they are organized at more complex levels as molecules, macromolecules, aggregations of these as minerals, planetary and astronomical bodies and eventually as life forms.

2.2 The elucidation of the relations between energy and matter has been an important feature of 20th century physics, leading to the conclusion that they are interconvertible. So elementary particles, for example, may be seen sometimes as forms of energy and sometimes as particles of matter. Nevertheless, a distinction between energy and matter is one reflecting a basic human sensory perception of the concepts and this makes a comprehensible and practical basis for their linear ordering. This is also consistent with a helpful indexing principle formalized by Ranganathan as a filing

2.3 The boundaries of the physics class are determined largely by gradation, which gives a broad sequence:

BA/BL Energy interactions and forms
BM/BQ Particles
BR/BW Bulk matter

Although particles are at the border between energy and matter they themselves reflect differences in mass and this principle provides the primary array within the particles facet. It also leads logically onto bulk matter, the forms of which are ordered primarily by increasing density.

2.4 The physical states of matter themselves form a disputed area which is sometimes claimed for physical chemistry. The term 'physical' is often used with a meaning broader than the one used here, which is of something pertaining to the world of energy and matter - ie to physics. So the forms themselves (gas, liquid, solid, etc) are preceded by the process of change of state, which is purely physical and does not involve chemical change.

2.5 A notable feature of the bulk matter class is that a number of the energy forms and interactions are manifested primarily and sometimes even uniquely in bulk matter. The principle of dependence in indexing rules that a class which depends entirely on the context of another class should be subordinated to the latter. As a result, the main enumeration of much of the electromagnetic phenomena class and of optics is given under bulk matter. Some of these concepts are dependenat on subjective human responses to the phenomena (eg luminosity and colour in optics) and this in turm implies the context of bulk matter. Other concepts reflect the problem of natural phenomena which are virtually defined by technological artefacts; examples of this are circuits and electric power.

3 Facets in Class B

The nature of these follows from consideration of the nature of facet structure summarized in sections 3/4 of the Introduction to AY/I:

3.1 Systems of macromatter, bulk matter

These consist primarily of the familiar physical states or forms of matter - gases, liquids, solids.

3.2 Parts, subsystems of macrosubstance

Although this includes concepts like surfaces, interfaces, etc. the most prominent are the elementary and other particles which make up the macroforms.

3.3 Processes and properties

This facet consists mainly of the different forms of energy and their interactions; it includes classical mechanics, gravity and the complete spectrum of electromagnetic radiation, including light.

3.4 General processes and properties

These could be regarded as part of Facet 3; but they are general in scope and not confined to physics. They are described in section 4.22 of the Introduction to AY.

3.5 Operations

These are the same concepts as described in section 4.3 of the Introduction to AY. But this facet in class B restricts them to their operation in physics.

3.6 Agents

These also are the same concepts are described in section 4.4 of the AY Introduction but restricted in their application to the field of physics.

3.7 Common subdivisions

These are described in the Introduction to AY (section 3.5).

4 Arrays in Class B

As noted in AY/I section 3.3, arrays reflect specific principles of division applied to the classes within a facet. For example, the particles in facet 4.2 are divided by Quantum property (to give conserved particles, etc), by Source (to give accelerator particles, etc), by Lifetime (to give stable particles, resonances, etc) and so on. At the more specific level of fluid flow (say), the schedule distinguishes more than seventeen different arrays defining types of flows.

5 Citation order in Class B

5.1 The citation order between the facets in physics is their order in 4.1-7 above. This reflects standard citation order. Bulk matter (BR/BW) represents the containing systems within which the subsystems (BM/BQ) are found; both of these display processes and properties (B9/BL); all these may be investigated by the operations in B6/8 and these in turn may utilize the agents in B3/5 to assist them. Finally, all concepts may be qualified by the CSD in B2.

5.2 Citation order between arrays is always clear from the inverted filing order. It is largely pragmatic (see section 4.6 in the Introduction to AY/I) but sometimes reflects principles akin to the standard citation order. For example, in the subordination of energy forms like absorption or diffraction to a particular wave-length of the electomagnetic spectrum; or, when two particles interact, the subordination of the particle with less mass to the one with more.

6 Filing order

This is inverted (see section 6 of the Introduction to AY/I). So the first cited facet (bulk matter) files last, the second cited facet (subsystems, particles) files next to last, and so on. Similarly, within a facet the first cited array files last and the last cited one files first.

7 Alternatives in Class B

7.1 The most far-reaching provision is that for the location of physics-based technologies at BX/BY instead of in technology at U/V.

7.2 An example of a less far-reaching alternative is at BB8 B Theories of mechanics; those theories frequently referred to as theories of mechanics (eg classical mechanics, quantum mechanics) may be located here rather than in the preferred location at B8B, where they are interpreted as theories applying to all or most of physics.

7.3 Another example is at BWY P where chemical crystallography may be collocated with general crystallography rather than separately in chemistry.

8 Notation and classmark building in Class B

8.1 The general features of BC2 notation are explained in section 8 of the Introduction to AY/I. The notes here give in some detail how the general principles are applied and interpreted.

8.2 Normal retroactive notation, in which earlier classes may be added directly to a later one with the omission of the initial letter B, is common throughout BA/BM; eg

B	Physics
BAF	Energy interactions and forms
BF	Waves
BFT	Scattering
BJ	Magnetism
BM	Particles
	(Special processes/properties)
BMM B	Symmetry [first enumerated subclass]

The fact that the first class special to particles doesn't appear until BMM B means that all the earlier subclasses of B (BA/BL) may be added directly to BM (to qualify particles) without clashing with the subclasses special to BM; eg BMF T Particles - Scattering.

8.21 From BM onwards, compounding between classes in BM/BW displays more complex relations and these are handled by a number of different arrangements involving intercalators. But with only rare exceptions (always noted in the schedule) Classes BA/BL can continue to be added directly as in the example above.

8.3 In many cases, two (or even three) initial letters can be dropped in retroactive notation. This is particularly the case in BR/BW Bulk matter, where all classes have at least two initial letters in common (BR, BS, BT etc.); eg

BRL	Optics (in bulk matter)
BRL L	Luminosity
BRL M	Colour
BRL ML	Luminosity, brightness

In the last class, three initial letters (BRL) are dropped.

8.31 This notational facet (BRL) is an example of one which has examples of dropping three letters yet apparently has assigned classmarks to some enumerated subclasses which would not allow this; eg

BRL X	X-rays
BRL XU	Grenz rays [the first enumerated class]

So BRL U (say) could not be added directly to BRL X (to give BRL XU) because this would clash with the enumerated special subclasses of BRL X. This, howevever, is not inconsistent with dropping three letters throughout BRL; the classes BRL U/BRL Y form an array of mutually exclusive classes between which no compounding is possible (there is no such thing as visible X-rays, for example). So the letter U is available for the first special subclass under BRL X without breaking the rule which holds through the rest of the notational facet BRL.

8.32 A similar situation holds under BVH Electrical properties of solids; eg

BV	Solids
BVH I	Electrical properties
BVH K	Charge
BVH N	Electrostatics
BVH NK	Charge [dropping BVH]

8.4 Use of intercalators

8.41 These may be used to extend the range of subclasses in a notational facet; eg in BF

BF	Waves
BFA	* Add to BF letters A/DR following B; eg
BFA BB	Wave energy [from BBB Energy]

This allows the enumeration of particular wave properties to begin at BFB rather than BFF (which would be necessary if direct addition of classes B/BE were allowed).

8.411 But compounds in BF formed by retroactive synthesis continue to be possible to the very end of the long list of properties (at BFV). So a further intercalator is inserted for classes BFG/BFY; all these may be divided retroactively by direct addition of A/F (following BF in BFA/BFF), but addition of classmarks from BG/BX need an intercalator to introduce them; eg

BFD	Frequency
BFL	Absorption
BFS	Collisions
BFS D	Frequency [by direct addition]
BFS GL	Absorption [by addition after intercalator]

8.412 A similar situation is found in BR/BW; generally speaking (there are some exceptions, always noted in the schedule) all these classes can add A/L (from BA/BL) directly but use M as an intercalator when adding classes from BM/BQ; eg BRM O Nuclear physics of bulk matter (where BR is Bulk matter and BO is Nuclear physics).

8.5 A more extended use of intercalators is found in handling complex relations in which concepts take on different roles, demanding varying citation orders which cannot be handled by direct retroactive notation. For example, at BNB instructions are given for the subdivision of any given particle in BM/BQ. Whilst direct addition of classes BMA/BMQ is allowed (dropping the two initial letters BM) beyond that, several distinct intercalators are use to introduce various types of compounds. So whilst Q introduces interactions with another particle, R and S introduce specification by another particle; eg BPQ P Electron-atom interaction (where BP is Atom and BNP is Electron) but BPQ RHW Neutral atoms (where BNH W is Neutral particles).

8.51 Direct retroactive notation may still be used within this framework of intercalators; eg

BNF	Anti-particles
BNN	Muons, mu-mesons
BNO	Neutrinos
BNO RN	Muon neutrinos
BNO RNR F	Muon anti-neutrinos

8.52 Another example of retroactive synthesis being effected by intercalators is found when compounding between different states of matter according to their (physical) phase relations. This maintains a consistent rule that the denser state is cited first in such compounds (and in filing order the denser state files after the less dense); eg

BV	Solids
BVU	(By state dispersed in the solid)
BVU N	Liquids in solid dispersions

It may be noted that the N for liquids is not a mnemonic got by using an earlier classmark (Liquids in general is BU). Maintenance of mnemonics would only be possible here at the expense of inordinately long classmarks. BC2 prefers brevity.

8.6 Interruption of normal synthesis A situation sometimes arises where one class demands a large expansion of an earlier class to meet its special needs. To squeeze this expansion into the earlier, more general classmarks would produce inordinately long classmarks; so the normal provision for retroactive notation is interrupted; eg

BS	Fluids
BSB	Fluid flow (from BB Mechanics)
BSF	Waves (from BF Waves)
BSG E	(Special attributes & elements)
	* Normal retroactive notation is interrupted here & resumed at BSM.
BSG P	Boundary layer
BSK Q	Flow around 3-d bodies
BSL T	Blunt bodies
BSM	(Other energy interactions)
	* Normal retroactive notation is resumed; Add to BSM letters GF/L following BR; eg BSM GH
BSM GH	Acoustic properties in fluid flow

8.61 In this way, the special needs of the fluids class is met whilst maintaining the consistent order of the general bulk matter class. There are numerous situations akin to this in class B; all such situations are clearly guided by explicit instructions for synthesis. They demonstrate the claim made in section 7.4 of the Introduction to AY/I that notation is always the servant of the order, not its master.

9 Some special problems in Class B

So far as is feasible, problems of definition and analysis and the practical decisions taken to meet them are noted in the schedules at the point of application. Others are implicit to some extent in the provision of alternatives. But a number of theoretical problems seem to call for a slightly extended explanation.

9.1 The problem raised by the overlapping of natural phenomena (the legitimate province of physics per se) with techniques of enquiry and other applications which extend the scale of some phenomena for practical purposes has already been noted (in section 1.42 of the Introduction to AY/I). The problem is particularly acute in physics, although the developing field of genetics technology is likely to equal it. Prominent examples are the techniques of stimulated emission of electomagnetic radiation in masers and lasers and the production of stereoscopic images in holography; but the whole area of electronics is full of examples of similar situations.

9.11 Two distinct questions are posed. Firstly, how far should the processes and properties developed be provided for as part of the naturally occurring phenomena they are based on? Secondly, should the general class for the resulting techniques and instruments be regarded as primarily dependent on the natural phenomena and therefore be subordinated to these within physics proper, or should they be treated as being primarily scientific and technical aids and therefore be located either in B or AY or in U/V Technology (according to the range of their application)?

9.12 It is not claimed that the solutions given in these schedules are ideal. but the provisions for synthesis in BC2 at least ensure that the concepts involved find a helpful place whatever the relationship is in which they occur.

9.121 When the class developed is almost entirely humanly devised and controlled it is preferred in technology; eg the large literature on electrical circuits and power production will be found mainly in technology rather than at BVH, as will that on semi-conductors (rather than at BVI). This should become clearer when the BC2 Technology class is published (in 1999 it is hoped).

9.122 When the techniques or instruments are developed primarily in the investigation of the natural phenomena they are collocated with that (whilst still appearing, of course, in the Operations and Agents facets too). Examples of this are at BRG V where thermometry and thermometers appear under the thermal properties of bulk matter, and at BRL where optical techniques and instruments are collocated with optics. But masers and lasers are located under techniques at B6K QM and holography at B7K.

9.2 Thermodynamics raises the question of whether it should be subordinated to bulk matter (at BRG P Thermal properties) or located at a more generally applicable position. In terms of the inverted schedule it seems clear that the fundamental thermodynamic laws should appear quite early, since they are applicable to all physical phenomena, including particle physics. So the general class is preferred at the beginning of the energy interactions and forms facet (at BAF).

9.3 A similar question as to the generality of concepts is found in the treatment of major bodies of theory (classical, relativistic, quantum, etc). Although the terminology used often suggests a scope less than the entirety of physics (eg classical mechanics, quantum mechanics) these have been put at B8B as theories applying to all physics.

9.4 A thorny problem of terminology arises over the distinctions between oscillation, vibration, wave motions and radiation Physics textbooks use these terms somewhat indiscriminately and numerous glossaries and dictionaries are little better. A controlled vocabulary like that of BC2 has to be more precise and it was decided to provide separate classes for all these terms, with a warning as to their ambiguity.

Enumeration of detailed subclasses is confined to one of them (Waves, at BF); within BR/BW Bulk matter the separate class for vibration and oscillation together (BE) is used as being more appropriate but it is divided in exactly the same way as is BF.

9.5 The location of the general classes for electrical energy and for optics raises the question of dependence. The latter principle requires many of the concepts involved to be subordinated to bulk matter (eg circuits, electrical power, luminosity, colour). In practice, most libraries probably prefer to keep the general classes on electricity and optics intact and not distributed between general physics and bulk matter physics. This is an example of literary warrant running counter to the theoretical requirements of consistent class formation. No perfect solution is possible and BC2 prefers the theoretical solution, since this is likely to be most helpful in the long run.

9.6 Particle physics poses the problem of a rapidly developing subject. The late appearance of this class of BC2 has at least the advantage of there having been a significant stabilization of this field over the last two decades or so. The definitions accompanying many of the terms attempt to indicate the underlying factors determining the treatment of the class.

9.7 A theoretical weakness of the Particles schedule, it may be thought, is the treatment of the basic interactions. The general concept of basic interactions (BMN V) naturally precedes the particles themselves, which are quasi-material entities. The notion of exchange particles (BMO) is subordinated to the general class as agents of the interactions. Theoretically, the individual particles would similarly be subordinated to the basic interaction they display. This is done for the gravitational force (BMP G) and the electroweak (BMP J); but the electromagnetic forces and the strong interactions have been subordinated to the types of particles which define them (charged particles and hadrons respectively) following the principle of dependence.

9.8 Another theoretical problem relating to particles is the appearance of gravity and electromagnetism as major energy interactions in general physics, ahead of their appearance as basic interactions. This separation reflects the fact that these interactions appear at every level of physics and not simply at the level of particles (where, in fact, gravitation effects are extremely difficult to find). On the other hand, the separation of the particles per se from the general interactions is consistent with the theory of gradation.

10 Practical classification in Class B

The general rules for classifying documents by BC2 are considered in the Introduction to BC2 (Section 7) and only a summary of the main points is given here.

10.1 Decide first which alternatives are to be used (if any) and delete clearly from the schedules those which are rejected. Once this is done it is done for good and henceforth the alternative are irrelevant to the library concerned. It also results in some simplification of the schedule.

10.2 Classification deals with the concepts which define the subject of the document and not with the terms used by the author to present the concepts (and which not infrequently may obscure the sense of some of the concepts). The schedule should provide a firm and predictable framework within which the classifier can locate the class (frequently a compound one) which accurately summarizes the overall subject of the document (as distinct from any subsidiary subjects referred to in the document).

10.21 In formulating the concept analyis of the document it is helpful to ask questions in a systematic order, posed in terms of the facets in their citation order. For example, does the subject of the document imply a context of bulk matter? If so, is it a particular state of matter? If so, does it deal with the relations of that form with particle physics and if so is it a particular particle? And so on.

10.22 The result of this systematic procedure is a summary statement of the document's subject in the form of a chain of terms; eg Fluids - High speed flow - Turbulent - Boundary layer. When the chain is complete, the classmark is built from an examination of the schedules, a consideration of the basic rules of retroactive notation and the noting of any special instructions along the way (in the use of intercalators, for example).

10.3 A few titles from recent publishers' catalogues in physics will demonstrate such chains and the building of the classmark:

10.31 Title: Detection of gravitational waves

Analysis of concepts in terms of the citation order, taking terms filing later before terms filing earlier gives the chain:

BGR	Gravitation
BGR F	Waves
BGR F74 J	Detecting

The first qualification of gravitation uses simple retroactive notation in which the earlier class BF Waves is added directly, dropping the initial B. The second qualification is also by direct addition of B74 J, dropping the initial B. The last classmark had itself been constructed by adding to B the numbers and letters following AY, the general class for detecting as a technique in scientific enquiry.

10.32 Title: The interacting boson-fermion model

Concept analysis produces the chain:

B	Particle physics
BNK	Bosons
BNK Q	(Interactions with other particles)
BNK QJ	Fermions
BNK QJ8 B	Theoretical models

The first qualification of bosons is by direct addition of the class BMQ Interactions between specific particles, dropping the first two letters BM. The authority for doing this is found in the instructions at BNB for the subdivision of any given particle. The Q in BMQ acts as an intercalator which introduces, with direct addition, the clasmark for all particles in BNB/BNY. Since Fermions is BNJ, only the J is added to BNK Q. The last addition is from B8B Theory & theoretical models in physics - itself taken from AY8 B Scientific theories & models.

10.33 Title: Turbulent boundary layers in supersonic flow

Concept analysis begins here with fluids in general; although supersonic flow more usually implies aerodynamic flow, there is nothing in the catalogue description to limit this study to gases. However, if examination of the document showed this to be the case, the commencing class would be BT (Gases) - nothing else would change.

BS	Fluids
BSJ T	High-speed flow
BSJ W	Supersonic flow
BSJ WJ	Turbulent flow
BSJ WJG P	Boundary layers

The schedule at BSJ, containing the array (Flows by speed) stops at BSJ T High speed flow. But the instruction at BSG E (which introduces a special expansion of the class of fluid flow) states clearly that the details given under gases in BT may be used anywhere in the fluids class if necessary; so the W in BSJ W is taken from BTJ W. The next two additions from the same set of arrays of fluid flows (BSJ Turbulent, and BSG P Boundary layers) are both added directly, dropping the two letters BS which are common to all the subclasses in BS.

10.34 Title: Quantized vortices in Helium II

The annotation states that the subject relates to superfluids, so the concept analysis begins with that as the context:

BS	Fluids
BSM PBF	Superfluids, helium II (superfluids)
BSM PBF DP	Vortices
BSM PBF DP8N	Quantized

The starting class Superfluids is enumerated in the schedule but is clealy the result of synthesis, locating it under atomic energy states in fluids. After that, simple retroactive notation gives Vortices by dropping the B in BDP and then Quantized by dropping the B in B8N (under quantum theory). In this context, helium II is interpreted as synonymous with superfluids, since no other superfluid is known. A more general work on helium II would go under atom physics, at BPU HH. The connection between the general class and this special aspect of it is made clear in the A/Z index, where the entries for the two classes come together under the term helium.

10.35 Title: Atomic transport in solids

Building from the concept appearing latest in the schedule gives the chain:

BV	Solids
BVM	(Relations with particles)
BVM NP	Atoms
BVM NPA T	Transport

Summary Outline of Sciences

Class AY/I. General science and the sciences. This is an inverted schedule: each class may be qualified by all preceding classes so far as they apply — e.g. History of scientific instruments AY4 27; Physics in earth science DGB

AY1 Science & technology (general)
* For technology, see U/V.

AY2 Science in general
(Common subdivisions)
AY27 History of science
AY29A Social aspects
AY2A Philosophy of science
AY2M Mathematics in science

(Operations & agents)
AY32 Research operations (general)
AY36 Practical scientific work
AY3B Equipment & materials
AY3U Equipment, plant
AY3X Laboratories
AY4 Instruments

AY62 Investigative techniques
AY69 Physical methods
AY6H Electromagnetic...
AY6KQ Microwave techniques
AY6X Chemical techniques
(By action on phenomena)
AY74 Detecting... Indicating...
AY76 Measurement.. Simulating...
AY7H Visualizing & imaging
AY7J Microscopy...
AY7M Spectroscopy

(Types of research)
AY7XN Non-experimental
AY82 Experimental (general)
AY8B Scientific theory

(General processes & properties)
* For qualification of individual sciences. Classes AY92/AY95 are not used here.
AY92D Distribution... Variation...
AY94 Conditions & parameters...
AY9B Dimension... Time... Space...
AY9G Systems characteristics
* As AYG/AYS; eg
AY9NP Continuous... Adaptive ...

(Subjects of scientific enquiry)
AYE The Sciences (in aggregate)
AYF General phenomena
* As Class 4 Phenomena.
AYG Systems... Complex phenomena...
AYNP Adaptive systems...

AYY Physical sciences
B Physics
C Chemistry
D Astronomy
DG Earth sciences
DU Geography
DY Ecology & environment
* Alternative to GX.

E Biology
EK Microorganisms
F Plants, botany
G Animals, zoology

GRX Applied biology
GS Agriculture
GT Animal husbandry
GU Veterinary science
GV Forestry
GX Human ecology
H Human biology... Medicine
I Human psychology
IRF Psychiatry

Summary Outline of Physics

Class B. Physics. This is an inverted schedule: each class may be qualified by all preceding classes so far as they apply — e.g. Particles — Waves BMF.

B Physics
B2 (Common subdivisions)
B2M Mathematics & statistics in physics

(Operations & agents)
B32 Research (general)... Methodology...
B36 Practical physics
B3B Equipment & materials
B3BD (Operations on)
B3U Equipment, plant
B4 Instruments, instrumentation
Investigative techniques
B69 Physical methods
B6H Electromagnetic... Electronics...
B6RGH Acoustic... Thermal...
B76 Measurement... Testing...
B7H Visualizing & imaging
B7J Microscopy... Holography...
B7M Spectroscopy...
B7X Vacuum techniques...
B8B Theoretical physics
B8D Relativistic... Quantum...
(General processes & properties)
B92 Distribution... Dimensions...
B9G Systems characteristics...
BAF Energy interactions & forms
BAG Thermodynamics (general)
BB Mechanics
BBB Energy... Forces...
BCH Statics...
BCX Dynamics
BDA Kinematics... Kinetics
BDS Periodic motion... Harmonics...
BE Oscillation & vibration
BF Waves
BFL Refraction...
(Special energy forms)
BGR Gravitation... Ballistics...
BGY Electricity & magnetism
BK Electro-magnetic radiation
(By frequency & wavelength)
BKM Radiofrequency...
BL Optics... Ultra-violet...

BM Particle physics
BM75 Detecting...
BM7T Acceleration...
BMBD Energy levels...
BMFS Collision... Scattering...
BMMD Quantum number properties
BMME Parity... Spin...
BMNV Basic interactions
* For Electromagnetic see BNG; For strong interactions see BNQP
BMPG Gravitational... Electroweak...
BNB (Types of particles)
BNFX Charged
BNG Electomagnetic interactions
BNJ Fermions... Boson...
BNM Leptons
BNQ Hadrons
BNQP Strong interactions
BO Nuclei, nuclear physics
BOW Fission... Fusion...
BP Atoms, atomic physics
BPBD Energy levels
BQ Molecules, molecular physics
BQU Ions

BQV Vacuums
BR Bulk matter
BRE Oscillation, vibration
BRGH Sound, acoustics
BRGP Thermal properties
BRH Electromagnetism
BRL Optics
* The main schedule is here.
BRM Particle physics in bulk matter
BRN States of matter
BRNP Change of state
BRST Dispersions... Colloids...
BRV Plasmas
BS Fluids
BSB Flow...
BSNP Change of state
BT Gases
BTX Condensed matter
BTXBF Energy bands
BU Liquids
BV Solids
BVH Electromagnetic properties
* The main schedule is here.
BW Crystallography

Outline of Science & Technology in general

AY1 Science & technology in general
. * Add to AY1 numbers & letters 2/9,A/V following AY;
. eg
AY1 29A . Social aspects of science & technology
29U . . Standards & standardization (general)

AY2 . Science in general
. . * For Technology in general, see U.
. . *Common subdivisions*
. . . * Add to AY2 numbers 2/9 from Auxiliary
. . . Schedule 1; eg
AY2 6C . . . Research (general)
7 . . . History of science
7C . . . . *By place*
8 . . . . *By period*
8C . . . . . Ancient science
8F . . . . . Medieval science
8H . . . . . Modern science
92 . . . Biography of scientists
9A . . . Social aspects of science, science & society
9EP . . . . Science policy
9X . . . Science as a discipline
A . . Philosophy of science
M . . Mathematics in science... Statistics...
Y . . *Relations between sciences*
. . *Operations & agents*
YQ . . . Organization & management of scientific
. . . work
. . . . * As TQ (including operations research).
AY3 2 . . . Research operations (general)
2C . . . . Methodology...
4 . . . Theory
. . . . * Alternative (not recommended) to locating after
. . . . Practical science, at AY8 B.
6 . . . Practical scientific work
7 . . . . Unwanted effects... Safety &
. . . . protection...
. . . . *Agents*
B . . . . . Equipment & materials (together)
. . . . . . *Operations on equipment & materials*
C . . . . Handling techniques...
F . . . . . . *Processes in equipment & materials*
G . . . . Deterioration...
J . . . . . . *Properties of equipment & materials*
JE . . . . Reliability... Responsiveness...
K . . . . *By energy interactions & forms*
. . . . * Add to AY3 K letters A/W following B
. . . . Physics; eg magnetic properties AY3 KJ.
. . . . . . *Parts*
NB . . . . Surfaces
. . . . . . *Types of equipment & materials*
PV . . . . Smart equipment & materials...
R . . . . . . Materials in general
RC . . . . Materials handling
RP . . . . *Types of materials*
U . . . . . . Equipment, plant

[Science & technology in general AY1]
[Science in general AY2]
[Operations & agents]
[Practical scientific work AY36]
[Agents]
. [Equipment & materials together AY3B]
. . [Equipment AY3U]

AY3 UR . . . Materials of equipment
UU . . . *Parts, components of equipment*
. . . . * As AY4.
. . . *Types by various characteristics*
UVT G . . . . Glassware...
X . . . Laboratories, scientific stations
AY4 . . . Instruments, instrumentation
. . . . Unwanted effects
AY4 37 . . . . . Interference...
3B . . . . *Operations on*
3J . . . . *Properties of*
3R . . . . *Materials of*
5 . . . . *Parts, components*
. . . . . *Parts by energy system*
. . . . . . * Add from Technology U/V
. . . . . . (notation provisional).
AC . . . . . . Electrical & electronic
. . . . . . components...
D . . . . . . Lasers...
F . . . . . . Optical components
G . . . . . . Thermal components...
H . . . . . . Mechanical components...
. . . . . *Parts by internal function*
K . . . . . . Switching devices...
L . . . . Transducers...
N . . . . . . Indicators...
R . . . . Recording devices...
S . . . . . . Input devices
SL . . . . Transducers...
SS . . . . Sensors...
T . . . . . . Output devices
TRY . . . . Viewing & display devices...
. . . . *Types of instruments*
AY5 5 . . . . . *By possession of a particular*
. . . . . *component*
. . . . . . * As AY4.
D . . . . . . Laser-driven instruments
HS . . . . . . Fluidic instruments
V . . . . . *Special to a given context*

Operations in scientific investigation
AY6 2 Investigative techniques
. *Serving all investigative objectives*
3 . . Data handling & processing...
. *By scale*
7 . . Microtechniques...
9 . Physical methods
. . * As BA/BW; eg
B . . Mechanical... Electro-magnetic...
KQ . . Microwave & optical techniques

Outline of Science & Technology in general

[Science & technology in general AY1]
[Science in general AY2]
[Operations & agents]
. [Practical scientific work AY36]
. . [Investigative techniques AY62]
. . . [Physical methods AY69]
. . . . [Microwave & optical techniques AY6KQ]

AY6 KS . . . . . Lasers...
LWY . . . . Radiological techniques... X-rays...
M . . . . Particle physics techniques
R . . . . Bulk matter techniques
UCW . . . . . Hydrodynamic technicques...
X . . . . Chemical techniques... Chemical
. . . . analysis...
. . . *By action on the phenomena investigated*
AY7 3 . . . . Production techniques...
4 . . . . Observing... Detecting... Indicating...
6 . . . . Measurement... Metric system... SI...
A . . . . Testing & evaluation... Standards...
F . . . . Simulation, modelling (simulation)
GY . . . . Visualizing & imaging (together)
I . . . . . Imaging
IX . . . . . Magnification, amplification...
J . . . . . Microscopy
K . . . . . Holography... Radiography...
M . . . . Spectroscopy... Spectrometry...
P . . . . Tracer techniques...
. . *Research by persons as agents*
WD . . . Individual research... Group...
. . *Research by environment*
WL . . . Special environments
X . . . . Vacuums (research environments)
XE . . . . Subsurface... Submarine... Space...
XN . . Non-experimental research
XU . . . Surveys... Expeditions... Fieldwork...
AY8 2 . . Experimental research (general)
2D . . . Design of experiments
. . *By broad objective*
5 . . . Fundamental research... Oriented...
B . . Scientific theory... theoretical models...
. . . * Alternative (not recommended) is AY3 4.
C . . . Hypotheses... Concept formation...
D . . . Theories special to a subject

AY9 *General processes & properties*
. * 9 is reserved here for use as a facet indicator to
. introduce the concepts below under the individual
. sciences. The classes AY9 2/AY9 Y are not used on
. their own at this position.
AY9 2D . Distribution... Variation...
4 . *Conditions & parameters*
4C . . Critical point... Volume conditions...
4J . . Pressure conditions...
B . . Dimension...
C . . . Time... Frequency...
D . . . Space... Mean free path...
DF . . . . One-dimension... Distance...

[Science & technology in general AY1]
[Science in general AY2]
[General processes & properties AY9]
. [Conditions & parameters AY94]
. . [Dimension... AY9B]
. . . [Space... Mean free path... AY9D]
. . . . [One-dimension... Distance... AY9DF]

AY9 DP . . . . Two dimensions...
G . Systems characteristics
. . * As AYG/AYS; eg
KV . . Continuous... Discontinuous...
LR . . Linear... Non-linear...

Subjects of scientific enquiry
AYE . The Sciences (in aggregate)... Nature...
. *By phenomena investigated*
AYF . . General phenomena
. . . * Add from Class 3 Phenomena.
AYF X . . Systems, communication & control, complex
. . phenomena
AYG . . . Systems, systemology
AYG 3D . . . . General systems theory
C . . . . Systems behaviour... State of system...
AYH . . . . . Variabiity... Change of state...
AYH W . . . . . Instability... Chaos...
AYI H . . . . . Adaptive behaviour...
. . . . *Subsystems*
AYJ . . . . . Structure of systems... Networks...
. . . . *Types of systems*
AYK B . . . . . Abstract... Stochastic... Open..
V . . . . . Continuous... Discontinuous...
AYL R . . . . . Linear... Non-linear... Reactive...
AYN J . . . . . Reactive...
MS . . . . . Dynamic... Stable...
P . . . . . Adaptive...
AYP M . . . . . Communication & control (systems
. . . . . theory)
N . . . . . . Information theory
AYQ . . . . . . Cybernetics...
AYS . . . . . . Artificial intelligence
. . . . * Alternative (not recommended) to locating
. . . . under computer science 8.
AYY . . Physical sciences

AY1 Science & technology (in general)
. * The connections of science with technology as a whole are very close, both conceptually and bibliographically. In some contexts (eg their history) the two subjects are so inextricably bound up at the general level that it is doubtfully helpful to maintain two separate classes. So this class (AY1) should take only those works which deliberately distinguish science and technology together from science alone or technology alone. When in doubt, prefer AY2/AYY.
. * Works which deal with technology alone go in class U/V.
. * Add to AY1 numbers and letters 2/9,A/V following AY so far as applicable; eg
AY1 29A . Social aspects of science & technology
29T Q . . Economic organization & management of science & technology
. . . * In society in general, at national or international level.
29U . . . Standards & standardization
. . . . * Agreement on and enforcement of standards by the scientific and technological communities.
. . . . * In cases of doubr, prefer Technology U29 U.
. . . . * For the operation of evaluation in arriving at standards, see AY7 E.
29U 295 D . . . . International standards organizations
29U 295 E . . . . National standards organizations
29U 29T Q . . . . Organization & management
. . . . . * Of institutions concerned with standards.
29U B . . . . . Standardization
. . . . . . * Process of agreeing on standards.
3AF . Research & development in science & technology
. . * See also R & D in science AY8 6M

AY2 Science in general
. * Science in its broadest sense, including the 'formal sciences' of logic and mathematics as well as empirical sciences (natural and social) and those applications often described as 'sciences' (eg medical sciences, engineering sciences, information sciences). It excludes only the Humanities (philosophy, history, religion, arts, philology).
. * For general works on Science of science (formal science, abstract science, scientia narrowly), see AK.
. * For works on a particular science, see the appropriate class (eg B Physics, HM Medical science).

[Science in general AY2]

Common subdivisions
. * Add to AY2 numbers 2/9 in Auxiliary Schedule 1, with the modifications and additions indicated.
AY2 . *Forms of presentation*
. . *Physical forms*
AY2 2H . . . Graphic materials
. . . . * For production of pictorial matter, see AY5 MR.
2U . . . . Graphs
3A . . Dictionaries & encyclopedias
3G . . Serials
3MC . . Aids to study
3NK . . Tables (general)
3NL . . . Tables of constants
. . . . * For constants as a subject, see AY7 6C.
3NP . . . Conversion tables
3P . . . Technical data
4 . Persons in science, scientists
. . * In the community, in society; including works on scientists & technologists together. For organization of personnel in technical operations, see Management of scientific work AY2 YQ.
. . * See also Biography of scientists AY2 92; Social aspects of science AY2 9A.
4C . . Profession
. . . * Collective body of persons in the subject.
4CD . . . Status of scientists
. . *Types of persons in the subject*
. . . * For technologists, see U24.
. . . * Add to AY2 4 letters L/R following K in KL/KR; eg
. . . *By sex*
4NW . . . . Women scientists
4PD . . . *By ethnic group*
. . . *By occupational characteristics*
4TQ . . . . Amateurs in science
4UM . . . . Technicians in science, auxiliaries in science, laboratory assistants
4UN . . . . Consultants in science
5 . Organizations in science
5C . . Conferences
. . . * For meetings as a form of communication in science, see AY2 5LP W.
5D . . International bodies, intergovernmental bodies
. . . *By name*
. . . . * Arrange A/Z; eg Unesco AY2 5DU.
5E . . National bodies
. . . *By country*
. . . . * Add to AY2 5F letters D/Z from Schedule 2.
5J . . . Government bodies
. . . . * For official bodies. For inter-governmental bodies, see AY2 5D.
5JQ . . Learned societies

[Science in general AY2]
[Common subdivisions]
[Organizations in science AY25]
[Learned societies AY25JQ]

AY2 5L Communication & information in science
. * Add to AY2 5L letters F/PS following KE so far as applicable (with the slight amendments indicated).
5LG L . Diffusion of scientific information
5LI . Audience (scientific communication)
5LK . Semiology in science, signs in science, symbols in science
. . * See also Constants & units (measurement) AY7 6C
5LK O . . Codes, notations
. . . * For metric system, see Metrology AY7 7E.
5LO . . Language of science, terminology, nomenclature
. . . * Add details from Language & linguistics XA so far as applicable; notation is provisional.
5LO T . . . Scientific terms, technical terms
5LO V . . . Abbreviations
5LP . Media (scientific communication)
5LP Q . . Mass media (scientific communication)
5LP W . . Meetings (scientific communication)
5MB . Science writing, scientific authorship
. . * For creativity, see AY2 9XB.
5MC . . Publishing scientific writings
5ME . . Translation (science writing)
5MG . . Reporting scientific work
5MH . . . Scientific papers
5MJ . . . Scientific journals
. . . . * Treated as a subject. For the journals themselves, see AY2 3G.
5MK . . . Science books
5MN . . . Correspondence in science
5MP . . The Press (scientific communication)
5MR . Pictorial matter (scientific communication)
5MS . . Illustration in science
5MT . . Cartoons in science
5MV . Telecommunications (scientific information)
. . * For electronic information systems, see AY2 5VB.
5NR . . Broadcasting (scientific communication)
. . . * Add to AY2 5N letters R/Y following KE; eg
5NR V . . . Radio (scientific communication)
5NS . . . Television (scientific communication)
. *Other elements in scientific communication*
5NW Y . . Popularization in scientific communication
5NX H . . Publicity
5NX KJ . . Public image (of science)
5P . Information science in science, documentation in science
. . * Details are taken from Classes 3/9 (not yet published). Notation is provisional.

[Science in general AY2]
[Common subdivisions]
[Communication & information in science AY25L]
. [Information science in science AY25P]

AY2 5Q . . Computers in science
. . *Operations on information records*
5R . . . Information handling & processing
. . . . * Selecting, storing, indexing, retrieval.
5S . . . . Bibliography
. . *Types of records*
. . . * Add to AY2 5T letters E/X following 2 in Auxiliary Schedule 1.
. . . * Add to AY2 5U numbers & letters 3/9,A/X following 3 in Auxiliary Schedule 1.
5VA . . Information services, information collections
5VB . . . Computerized information services
5VD . . . . International information services
5VE . . . . National information services
5VF . . . . Regional information services
5VG . . . . Local information services
5VJ . . . . Publicly-owned information services
5VM . . . . Academic information services
5VP . . . . Private information services, profit-making information services
5VS . . . . Special information services
. . . . . * Restricted as to subject. General works only; services for a specific subject go with the subject.
5W . . . Libraries
. . . . * Add to AY2 5W letters D/S following AY2 5V; eg National science libraries AY2 5WE.
6A Education of scientists, study & teaching of science
. * Alternative (not recommended) is to locate in Class J Education.
. * Add to AY2 6A numbers & letters 2/9, A/Z following J.
6AJ LO . Laboratory demonstrations (science education)
6C Scientific research
. * By definition, all activity in a pure science is implicitly research, or investigation with a large research element. Theory is closely allied to practical and experimental work and the two activities have been integrated at AY3/AY8 below.
. * This general class for research is retained here primarily to allow qualification, if necessary, of classes preceding AY3 in filing order.
. * For scientific methodology broadly, see Science as a discipline AY2 9X; for the operations involved in practical research, see AY36/AY7.
6QT Exhibitions, museums
. * See also Physical models AY7 FYS

[Science in general AY2]
[Common subdivisions]
. [Exhibitions AY26QT]

AY2 6R . . *Science museums by place*
. . . * For special types of collections (eg of minerals) see subject concerned.
. . . * Add to AK6 R letters C/Z from Schedule 2.
6SV . Curiosities in science, mysteries in science
6SW . . Natural wonders (of the world)
. . . * Works restricted to the earth sciences go in D.
6SX . Frauds, deceptions, hoaxes
6SY . Recreations (scientific), popular science

6Y History & philosophy of science
7 . History of science
. . * Including works which consider technology as well but only insofar as it is contributory to, or inseparable from science. For history of technology alone, see U27.
. . * See also scientific revolutions AY2 9XQ
77 . . Historiography of science
. . . * History of writings & studies in history of science.
. . . * Notation provisonal; this class is taken from L History (not yet published 1998).
. . . *History of science as a discipline*
78 . . . . Sources and methodology
. . . . . * As a subject itself. Early scientific works (up to 1800) are classed under the period in which they were contemporary. If limited to a specific subject, they go under the history of that subject.
79 . . Ancilliary studies in history of science
. . *By period and place*
. . . * The universality of science is such that period is much more significant characteristic than place. So the preferred arrangemnt here is to cite period before place.
. . . * Citing period first raises serious problems in that the literature can (and does) select any of an infinite number of different periods within which it considers the history of science.
. . . * The solution below attempts an arrangement of optimum usefulness and clarity by following the citation order: Broad period- -Place- -Specific period. The broad periods are enumerated; the places follow Auxiliary Schedule 2; the specific periods may be taken from one of the three options in Auxiliary Schedule 4 (but 4A is the recommended one).
. . . * In addition to the above basic provision, it may be noted that any broad period, or any place, or any broad period qualified by place, may be further divided using 7 to introduce specific periods or 8 to introduce specific places. Examples (using Auxiliary 4A Periods) are AY2 8C7 EPN Ancient science 600/300 BC;

[Science in general AY2]
[History & philosophy of science AY26Y]
[History of science AY27]
[By period and place]

AY2 8CR 7EP N Ancient Greek science 500/001 BC: AY2 8J7 KKV 17th century science, 1640/1680; AY2 8J8 E7K KV Science in Britain 1640/1680.
* An alternative (not recommended) is provided at AY2 7B, allowing period to be cited after place at AY2 7D.

AY2 7A Early works (on history of science)
. * For works published before 1800 AD.
7B *By period*
. * Alternative (not recommended) for libraries citing place before period (see note above). If this option is taken, proceed as follows:
. * Add to AY2 7B letters from Auxiliary Schedule 4B (for Time); eg 18th century AY2 7BK.
. * These periods may then be used to qualify the places in AY2 7C; eg French science in the 18th century AY2 7F7 BK.
7C *By place*
. * For general works on history of science in a given place. If confined to a particular period, cite period first (from AY2 8 onwards).
. * Add to AY2 7 letters C/Z from Schedule 2.
By period
8A . *Periods from the earliest times*
. . * Add to AY2 8A letters D/Y following AB in Auxiliary Schedule 4B; eg
8AR . . Prehistory science, primitive science
. . . * For works on science in contemporary non-literate societies in general, see AY2 9KS LM. Individual non-literate societies go with the nation state in which they live.
8AY . . . Neolithic science
8B . . Ancient & medieval science (together)
. . . * Do not divide by specific period.
8B8 . . . *By place*
8C . . Ancient science (general)
8C7 . . . *Specific periods of ancient science*
. . . . * Add to AY2 8C7 letters DF/EU from Auxiliary Schedule 4A; eg AY2 8C7 EPN is Ancient science in 600/300 BC.
. . . *By place*
. . . . * Add to AY2 8C letters C/Y following C in Schedule 2 (for places in the ancient world); eg
8CD . . . . Mesopotamia (Ancient world), Fertile Crescent
8CD C . . . . . Babylonian science
8CL N . . . . Asia (ancient science)
8CL X . . . . Americas (ancient science)

Science & technology (in general)

[History & philosophy of science AY26Y]
[History of science AY27]
. . [By period]
. . . [Periods from the earliest times AY28A]
. . . . [Ancient science general AY28C]
. . . . . [By place]
. . . . . . [Americas ancient science AY28CLX]

AY2 8CL ZA . . . . Mayan science
8CM . . . . . . Africa (ancient science)
8CN . . . . Egypt (ancient science)
8CQ . . . . . . Europe ancient science
8CQ Y . . . . Classical antiquity (ancient science)
8CR . . . . Greece (ancient science)
8CY S . . . . Sicily (ancient science)
8D . . . Medieval and modern science (together)
. . . . * Do not divide by specific period; for this, see the individual broad periods Medieval, Renaissance, 17th century, etc.
8D8 . . . . *By place*
8E . . . . Medieval & Renaissance science (together)
. . . . . * Do not divide by specific period; see the broad periods Medieval and Renaissance.
8E8 . . . . . *By place*
8F . . . . . Medieval science (say 500/1350 AD)
8F7 . . . . . . *Specific periods in medieval science*
. . . . * Add to AY2 8F7 letter F in Auxiliary Schedule 4A.
8F8 . . . . . . *By place*
8G . . . . . Renaissance science (say 1350/1550)
. . . . . . *Specific periods in renaissance science*
. . . . * Add to AY2 8G letters G/HM in Schedule 4A.
8G8 . . . . . . *By place*
8H . . . . Modern science (1550 to date)
. . . . . * Do not divide by specific period. Use the individual centuries (17th onwards) for this, adding to each the relevant block of letters for starting dates (to each of which may be added the letter for the duration of the period); eg AY2 8LP FW History of science 1820/1850.
8H8 . . . . . *By place*
8J . . . . . 17th century
8K . . . . . 18th century
8L . . . . . 19th century
8N . . . . . 20th century
8P . . . . . 21st century
8V . . . . Trends in science, futurology in science
. . . . . * Speculations on future history of science.
. . . . . * For scientific forecasting, see AY2 9EF.
92 Biography of scientists
93 . Collective biography
94 . Individuals

[Science in general AY2]
[History & philosophy of science AY26Y]
[Biography of scientists AY292]
. [Individuals AY294]

. . * Arrange A/Z.
AY2 96C Progress in science
96D . Discovery & invention in science
. . * When treated as a result of science. If in doubt, prefer Technology U.
. . * For invention, patents, etc., see Technology U26 QH.
96E . . Discovery in science
. . . * The finding of something existing in nature but not hitherto known.
. . . * For scientific explanation, see AY2 9XL.
9A History & philosophy of science (together)
. * The preferred arrangement is to locate the philosophy of science at AY2 A. This location provides an alternative (not recommended) for libraries wishing (1) to keep together the history and philosophy of science in general, and (2) to subordinate to it the history and philosophy of the individual sciences. If this option is taken, proceed as follows:
. * Add to AY2 numbers 7/96 for history (as at present);
. * Add to AY2 9 letters A/L following A (for the philosophy of science);
. * Add to AY2 9M letters A/K from the main classification (for history & philosophy of the individual sciences).
. * Add to AY2 9N letters A/X following AY2 9 below (for Social aspects of science); eg AY2 9NE P Science policy.
. * Each individual science may then be qualified as follows (where the hyphen represents its classmark):
. * Add to -2 numbers 7/96 for history; eg History of physics AY2 9MB 27;
. * Add to -29 letters A/L for philosophy; eg Philosophy of physics AY2 9MB 29A
. * A further alternative (not recommended) is to add the social aspects of an individual science after its history and philosophy, using - 2M to introduce the divisions of AY2 9; eg Science policy relating to physics AY2 9MB 29M EP.

Social aspects of science

[Science in general AY2]
[History & philosophy of science AY26Y]
[History & philosophy of science together AY29A]

AY2 9A Social aspects of science, society & science
. * In much of this literature, science and technology are inextricably bound up. So all classes below are assumed to include technology whenever this is considered along with science; eg AY2 9EP includes works on science & technology policy together.
. * Add to AY2 9 letters E/J following K9.
. * Add to AY2 9K numbers & letters 9Q,A/Y following K at K9Q/KY.
. * A selection of concepts is given here for convenience.
9EF . Forecasting
9EP . Science policy
. . * See also Economics of science AY2 9T
. . Organizations
9EP 5 . . . Policy making bodies
9EP 5J . . . . Government
9G . . Planning and development in science
. . . * See also Research & development in science & technology AY1 3AF
9KA . . Social system of science, science system
. . . * Science as a subsystem of society as a whole.
9KB V . . . Scientific culture
9KC . . . Social processes in science
9KC E . . . . Science input to society
. . . . . * Science as agent of change, etc.
9KF D . . . Attitudes to science
. . . . * See also Antiscience AY2 9KI G
9KH K . . . Rewards in science
9KH KM . . . . Awards and medals in science
. Hostility
9KI G . . Antiscience
. Communities
9KL K . . Scientific community
9KL KFB . . . Institutionalization of science
. . . . Informal structures
9KL KPM . . . . . Invisible colleges
. . . . . . * For specific formal organizations in science, see Organizations AY2 5.
9KS LM . . Non-literate societies
9KW . Custom, folklore & mythology in science
9L . *Other aspects of society*
. . * Add to AY2 9 letters L/Y from the main classification; eg
9P . . Religion in science
. . . * Religious beliefs & loyalties in scientific work. For impact of science on religion, see Class P.
9PY . . Morality in science, ethics in science
. . . * See also Philosophy of science AY2 A
9PY MJ . . . Values in science
9PY NH . . . Responisibilities of science
9R . . Politics & science

[Science in general AY2]
[Social aspects of science AY29A]
[Other aspects of society AY29L]
[Politics & science AY29R]

AY2 9RA OF . Internationalism in science
9S Law & science
. * The law of special subjects is preferred in Class S. Locate here only those works dealing with legal aspects as they may impinge on scientific activity.
9SB LJ . Intellectual property in science
9SB M . . Patent law & science
9T Economics
9TC D . Growth & development
9TC DQ . . Needs of science (economics)
9TC DR . . . Science priorities
9TC DS . . Resources in science
9TP G . Modern economic systems & science
9TP H . . Less developed countries (science in), LDC science
9TP K . . Developing countries (science in)
9TQ . Economic organization & management of science
. . * In society in general, at national or international level; for the organization of scientific work per se (of research, experiments, etc.), see AY2 YQ
. . * For Standards & standardization, see AY1 29U.
9V Technology in science
. * General works on role of technology in science. For specific applications of technology in science (e.g. instrumentation) see application. For role of science in technology, see Technology (Class U).
9VN . Environmental technology in science

[Science in general AY2]
[Social aspects of science AY29A]
. [Other aspects of society AY29L]
. . [Technology in science AY29V]
. . . [Environmental technology in science AY29VN]

AY2 9X Science as a discipline, scientific method
. * General investigation of the aims, concepts and principles of reasoning in scientific enquiry and the relationships between its subdisciplines.
. * For methodology in the narrower sense of the study of the methods or procedures in science, see AY3 2C. For theory as a particular component of this, see AY8 B.
. *Attributes of scientific enquiry*
9X5 . . Autonomy of science
9X6 . . Constraints in science, limits of science
9X8 . . Criteria in science
9XA . . Non-rational elements in scientific enquiry
9XB . . . Imagination in science, creativity in science
9XC . . . Intuition in science
9XE . . . Aesthetics in science
9XG . . Interdisciplinary relations between sciences
9XJ . Objectives of science
9XL . . Explanation in science
9XM . . . Theories in science (as a subject), hypotheses, models in science (as a subject)
. . . . * As a subject of study. For theory as an investigative method in scientific research, see AY3 D.
9XP . . . . Paradigms
9XQ . . . . . Scientific revolutions
9XS . . . Scientific laws, laws of nature

A . Philosophy of science
. . * An alternative (not recommended) is to collocate the philosophy of science with the history of science, by locating it immediately after it, at AY2 9A (qv).
. . * Add to AY2 letters A/L following A in AA/AK; eg
BA . . *Viewpoints, doctrines*
CKQ . . . Evolutionism in science
GG . . Metaphysics
GQD . . . Causation in scientific explanation
GR . . Epistemology
GSV . . . Scientism
HK . . *Ethics*
L . Logic in science
. . * See also scientific method AY3 3F

[Science in general AY2]
[Science as a discipline AY29X]
. [Logic in science AY2L]

AY2 M . Mathematics in science
. . * General applications only. For applications of mathematics to specific topics in science, see the topic.
. . * For mathematical models as types of theory, see AY8 B2M.
. . * Add to AY2 letters M/X following A in AM/AX; eg
M8L . . Functions, operators
M8N . . . Named functions
. . . . * Arrange A/Z.
M9L . . Equations
M9N . . . Named equations
. . . . * For inequalities, use 2M9 O.
. . . . * Arrange A/Z.
P2 . . Mathematicians, names of scientists (as principles, etc)
. . . * For systems, principles, laws, etc referred to by the name and connoting concepts not easily defined by a more specific subject; eg BB2 P2H Hamiltonian system.
QK . Coefficients
X . Statistics in science
XG . . Statistical probability
XHB B . . . Expected values, errors
XS . . Statistical analysis, data analysis (statistical)
Y . *Relations to other sciences*
. . * Add to AY2 Y letters B/I in the main classification; eg relations to astronomy AY2 YDA

Operations & agents
. *Preliminary operations*
YMH . . Preparing research proposals
YMJ . . Planning & design of research
YMP . . Disseminating & publishing research reports
. . . * See AY2 5MB
YQ . Organization & management of scientific work
. . * One subject of some importance in science (operational research) is given below, together with some major classes from the detailed schedule at TQ which gives operational research its context.
. . * Add to AY2 Y letters Q/V following T.
YQJ . . Management services
YQK . . . Communication & control (scientific work), systems analysis in scientific work
. . . . * Its application to scientific work; as an object of study itself, see AYD.

[Science in general AY2]
[Operations & agents]
[Organization & management of scientific work AY2YQ]
[Management services AY2YQJ]
[Communication & control scientific work AY2YQK]

AY2 YQR Management techniques, scientific management
YQS . Operational research, operations research, OR
. . * Improvement of efficiency of management by techniques derived from numerical analysis, q.v. (AX7 6D). Aims to increase the results/effort ratio by increasing the first or decreasing the second without altering the other.
YQT B . . Problem-solving techniques
YQT L . . . Mathematical & statistical techniques in OR
. . . . * Add to AY2 YQT letters M/X following A in AM/AX.
YQT M . . . . Mathematical models in OR
YQU C . . . Decision theory
YQU D . . . . Decision analysis
YQU E . . . . . Decision trees
YQU F . . . . . . Branch & bound analysis
YQU H . . . . Risk analysis
YQU K . . . Optimization (OR)
YQU L . . . . Mathematical programming
YQU M . . . . . Linear programming
YQU N . . . . . . Simplex method
YQU P . . . . . . Transportation method (OR)
YQU PW . . . . . . Assignment method
YQU Q . . . . . Non-linear programming, higher order programming
YQU QV . . . . . . Quadratic programming
YQU QX . . . . . . Convex programming
YQU S . . . . . Static programming
YQU T . . . . . Dynamic programming, control programming
. . . . . . * See also Inventory problems AY2 YQW S
YQU V . . . . . Integer programming
YQV B . . . Simulation (operational research)
YQV C . . . . Computer simulation (operational research)
YQV F . . . Heuristic programming
YQV J . . . Network analysis, network planning, project planning
. . . . * See also Networks (systems theory) AYJ W
YQV K . . . . Critical path scheduling, critical path analysis
. . . . . * See also Branch & bound analysis AY2 YQU F

[Organization & management of scientific work AY2YQ]
[Management services AY2YQJ]
. [Management techniques AY2YQR]
. . [Operational research AY2YQS]
. . . [Problem-solving techniques AY2YQTB]
. . . . [Network analysis AY2YQVJ]
. . . . . [Critical path scheduling AY2YQVK]

AY2 YQV M . . . . . . Critical path method, CPM (networks)
YQV P . . . . . . Programme evaluation & review technique, PERT (networks), programme evaluation procedure, PEP (networks)
YQV Q . . . . . . Line of balance networks
YQV R . . . . . . Other scheduling methods
. . . . * Arrange A/Z; eg WASP.
. . . . *Other special procedures*
YQV T . . . . . Action research
. . . *Operational problems*
YQW D . . . . Deterministic problems
YQW F . . . . Probabilistic problems
YQW G . . . . . Stochastic problems, randomness
YQW J . . . . . Allocation of resources
YQW L . . . . . Sequencing problems, scheduling problems
YQW N . . . . . Routing problems
YQW NUN . . . . . . Linear programming
YQW NUT . . . . . . Dynamic programming
YQW Q . . . . . Queuing problems, waiting time problems
. . . . . . Mathematics
YQW QTM . . . . Queuing theory
YQW QXD . . . . . . Delay problems
YQW S . . . . . Inventory problems
YQW V . . . . . Replacement problems
. . . . . . Statistics
YQW VTX . . . . Renewal theory
YQW X . . . . . Search problems
YQX B . . . . . Competitive problems
YQX BUC . . . . . . Decision theory
YQX BVB . . . . . . Simulation
YQX BVD . . . . Operational gaming
YRE . . Organization & methods, O & M (management)
YRF . . Work study
YRF H . . . Method study
YRK B Management functions
YRY Management resources
YSO . Office services, administrative services
YTS . Funding research in science
YU . Personnel management
YVE Production management
YVJ . Equipment & materials (production management)

[Science in general AY2]
[Operations & agents]
[Organization & management of scientific work AY2YQ]
. [Production management AY2YVE]
. . [Equipment & materials production management AY2YVJ]

. . . * For general works only, on management of equipment & materials. For specific equipment & materials as integral elements in practical scientific work, see AY3 B.

AY2 YVP . . Production planning & control
YVT . . . Quality control
. . . . * For Standards & standardization see AY1 29U.

AY3 2 Research operations (general)
. * It is often not feasible to distinguish between 'pure' and applied research or between research per se and general scientific work. So for specific methods, techniques, types of equipment, etc. only one schedule is provided, covering both research narrowly and scientific investigative procedures in general. For the latter, see AY6; for technological research in general, see U3.
. * Practical work, both in its operations and the equipment used, is essentially an agent of research (a form of investigation).
. * Theory per se, regarded as the exposition of the principles of the subject concerned, is intimately linked with the phenomena and concepts investigated; physics is a particularly clear example of this. For this reason, the preferred arrangement is to file theory after practice (observing the retroactive principle) thereby collocating it with the concepts making up the subject proper.
. * In some sciences, it may be thought that this relationship is less significant; so an alternative is provided at AY3 4 for libraries wishing to file theory before practice (reflecting the principle of increasing concreteness in filing order).

2B . Procedures & methods (research)
. . * For classification, see AY8 BT; for Scientism, see AY2 GSV; for techniques narrowly, see AY6 2.

2C . . Methodology
. . . * The study of research methods themselves, especially as to their logical bases. For methodology in the wider sense, defining the nature of science as a discipline, see AY2 9X.
. . . * Those concepts in this class which are special to the social sciences are given in K6/K9.
. . . * See also Philosophy of science AY2 A
. . . * Add to AY3 2 letters D/G following K6; eg

2D . . . Qualitative methods (general)
2E . . . Quantitative methods (general)
2F . . . Non-empirical methods
2G . . . Empirical methods
. . . . * Usually assumed. For induction, see AY8 BL; for hypothesis testing, see AY8 CJ.

[Science in general AY2]
[Operations & agents]
[Research operations general AY32]
. [Procedures & methods research AY32B]
. . [Methodology AY32C]
. . . [Empirical methods AY32G]

AY3 2L . . . Analytical methods
2N . . . Synthetic methods
2Q . . . Comparative investigations
2R . . . . Time comparison research
2S . . . . Longitudinal research
2T . . . . Secular studies, long-term studies
4 . . Scientific theory, theoretical models
. . . * Alternative (not recommended) to locating at AY8 B; If this option is taken proceed as follows:
. . . * Add to AY3 4 letters B/C following AY8; eg Hypotheses AY3 4C.

4D . . . Theories particular to a subject
. . . . * Alternative (not recommended) to locating after Practical scientific work, at AY8 D.
. . . . * For works under specific classes dealing with the theories per se.

5 . . Experimental research
. . . * Alternative (not recommended) to locating at AY8 2.
. . . * Add to AY3 5 letters A/Y following AY8 2; eg

5D . . . Design of experiments
6 Practical scientific work
. * The classes of procedures, techniques, equipment and instrumentation given below are usually considered in the context of experimentation. So the latter is included here, as being nearly synonymous with practical scientific work. Provision is made at AY8 2 for general works on experimental research should these distinguish it from practical science.
. * This location provides for truly general works on these classes and for the qualification of all other types of scientific enquiry so far as applicable; eg Field studies - Equipment & materials AY7 YF3 B.
. * Note that the building of classmarks for compound subjects will sometimes entail building forward (not retroactively, as is the general rule in BC2); eg Instruments - Calibration - Testing AY4 4BW 7A (when the general class Testing AY7 A files after the general class Instrument calibration AY4 4BW).
. * Building forward by direct addition is not always feasible. If the classmark added to ends in a numeral (eg Instruments AY4) direct addition may produce an ambiguous classmark since numerals may be used for its own enumerated subclasses. In such cases (which are rare) building forward is still possible, but an intercalator must introduce it; in these cases, proceed as follows (where the hyphen represents the classmark added to):
. * Add to -36 numbers 3/9 following AY.

[Science in general AY2]
[Operations & agents]
[Practical scientific work AY36]

AY3 7 Unwanted effects, hazards, accidents, safety & hygiene precautions
. * For Interference, see Instrumentation AY3 48K.
. * Add to AY3 7 letters A/Y following HJ; eg
7DV . Reporting systems (hazards), warning systems
7DW . Control measures (hazards)
7E . Protective measures, monitoring & protecting (hazards), safety
7ES . . Protective equipment
7EU . . . Protective clothing
7EV . . . . Specific items
. . . . . * Arrange A/Z; eg masks AY3 7EV M.
7EX . . . Special to a particular science
. . . . * Notation reserved for amplifying particular contexts.
7H . *Types of hazards*
7I . . Aberrations, anomalies
7J . . Pollution hazards in research
7K . . Radiation hazards in research
7KL . . . Ionizing radiation hazards
7LH . . . Electromagnetic radiation hazards
7LM . . . Particulate radiation hazards
7MG . . Dangerous substances, dangerous materials
7N . . . Specific substances
. . . . * Add to AY3 7N letters H/U following C.
8E . . Error, accidental error, human error
. . . * For error analysis, see Testing & evaluation AY7 DA.
8J . . Hazards special to a given context
. . . * Eg optical aberration; interference in instrumentation AY4 38K.
Unwanted effects special to instrumentation
. * See Instrumentation AY4 38K
8S Falsification of results
. * See also Error analysis (evaluation) AY7 DA

Agents
B . Equipment & materials (together)
B37 . . Unwanted effects
. . *Operations on equipment & materials*
. . . * Most of the literature relates specifically to equipment or to materials and the terms below will be used mainly to qualify these separately. This class (AY3B/AY3P) takes only those works covering materials, equipment and instrumentation together.
BD . . . Design
BF . . . Installation & use
BG . . . Workshop techniques
. . . . * Welding, machining, lubricating, etc.
BGP . . . . Coupling
. . . . . * Of one piece of apparatus to another.
BH . . . Maintenance

[Science in general AY2]
[Operations & agents]
[Practical scientific work AY36]
[Agents]
[Equipment & materials together AY3B]
[Operations on equipment & materials]
. [Maintenance AY3BH]

AY3 BK . Testing & monitoring (equipment & materials)
. . * Add to AY3 BK letters A/E following AY7; eg Measuring instruments - Standards AY7 643 BKE.
BN . Control of operations (on equipment & materials)
BR . Performance criteria control
. . * For performance criteria control instruments, see AY4 3BR.
. . * For calibration, etc., see Instruments AY4 4BW.
C . Handling techniques (general)
. . * For handling techniques special to a given science, see latter.
CD . . Synchronizing
CF . . Manipulating
. . . * For example, by glove boxes.
CH . . Holding, supporting
CI . . . *By agent*
. . . . * Arrange A/Z; eg clamps, mandrels, tripods.
CJ . . Joining, joints
CK . . Closing
CL . . . Materials for closing
CM . . . . Packing materials, caulking
CP . . . Equipment for closing
CQ . . . . Seals
CR . . . . Other
. . . . . * Arrange A/Z.
CS . . Fastening
CT . . Containing
CU . . Sequencing
CV . . Positioning
CW . . Lifting
D . . *Special to materials*
. . . * See Materials handling AY3 RC
Processes in equipment & materials
F . *Effects of physical phenomena on equipment*
. . * Add to AY3 F letters A/W following B; eg Effect of pressure AY3 FBJ.
G . Deterioration
Properties of equipment & materials
JC . Characteristics (equipment & materials)
JE . Reliability
JG . Accuracy
JI . Sensitivity
JK . Responsiveness

Practical scientific work

[Science in general AY2]
[Operations & agents]
[Practical scientific work AY36]
[Agents]
[Equipment & materials together AY3B]
[Properties of equipment & materials]
. [Responsiveness AY3JK]

AY3 JL . Gain
. . * See Instrumentation AY4 3JL
JS . Stability (equipment & materials)
JT . Time response
. . * See AY4 84J T
. *Physical properties*
. . * Add to AY3 K letters A/W following
. . B; eg
KGH . . Acoustic properties
KGP . . Thermal properties
KH . . Electromagnetic properties
KL . . Optical properties
L . . Chemical properties
. . . * Add to AY3 L letters following C
M . *Properties special to a material or*
. *device*
Parts
NB . Surfaces (equipment & materials)
Types of equipment & materials
PV . Intelligent equipment & materials,
. smart equipment & materials
. . * Equipment or material which responds
. . to external stimuli in a specific,
. . controlled way; eg lenses reacting to
. . sunlight. Used particularly when
. . properties demonstrate use of artificial
. . intelligence.

R Materials in general (in practical
science)
. * For materials technology in general, see
. Class U.
. * Add to AY3 R numbers & letters 6/P
. following AY3; eg materials reliability
. AY3 RJE.
. *Operations on materials*
RC . . Materials handling
RD . . . *Special to materials*
. . . . * Add to AY3 RD letters following
. . . . UF (Materials technology) so far as
. . . . applicable. Notation is provisional
. . . . until publication of Class U/V
. . . . Technology.
RDH . . . . Bulk handling of materials
. *Processes, properties & parts of*
. *materials*
. . * Add to AY3 R letters F/N following
. . AY3 (Equipment & materials); eg
RJE . . Reliability of materials
. *Types of materials*

[Science in general AY2]
[Operations & agents]
[Practical scientific work AY36]
[Agents]
[Equipment & materials together AY3B]
[Materials in general in practical science AY3R]
. [Types of materials]

AY3 RPV . . Intelligent materials, smart
. . materials
. . . * See definition at AY3 PD.
. . *By state of matter*
. . . * Add to AY3 R letters Q/W
. . . following B (with modifications; eg
RQV . . . Films
RQX . . . Coatings
RS . . . Fluids (materials)
RV . . . Solids (materials)
. . *By chemical constitution*
. . . * Add to AY3 S letters H/W following
. . . C; eg
SIB . . . Metals (materials)
SO . . . Organic materials (materials)
. . *Specific materials*
TF . . . Fibres
TG . . . Glass
TH . . . Ceramics
TJ . . . Plastics
. . *By utility*
TL . . . Fuels & explosives (practical
. . . work)
TN . . . Adhesives (practical work)
TP . . . Light sensitive materials
TR . . . Solvents (practical work)
TS . . . Insulating materials
TT . . . Refractories (practical work)

U Equipment & plant, equipment
(general), apparatus (general),
devices (equipment)
. * Add to AY3 U letters BD/M and R/T
. following AY3; eg
U37 . Unwanted effects
. *Operations on*
UBD . . Design
UCK . . Closing
. *Properties*
UJE . . Reliability
UR . *Materials of equipment*
UTS . . Insulating materials
UU . *Parts of equipment*
. . * Add to AY3 UU letters A/Y following
. . AY3 N; eg
UUB . . Surfaces
UV . *Types of equipment*
. . *By material*
. . . * Add to AY3 UV letters RP/V
. . . following AY3 so far as applicable; eg

[Operations & agents]
[Practical scientific work AY36]
[Agents]
[Equipment & materials together AY3B]
[Equipment & plant AY3U]
[Types of equipment AY3UV]
. [By material]

AY3 UVT G . Glassware
Types of equipment by function
. * See function; note that this may sometimes entail building forward, rather than retroactively; eg equipment for handling AY3 C3U.
V . Intelligent equipment, smart equipment
. . * For intelligent instruments specifically, see AY4 5V.
W *Types of equipment special to a subject*
X Laboratories, scientific stations
. * Regarded as aggregates of the total resources of equipment, materials, etc. in a given operation.
. *Personnel*
X24 UM . . Assistants, technicians
X3B D . Design & layout
XL . Buildings
XM . Mobile laboratories
XN . Testing laboratories
XP . Analytical laboratories
XR . Control laboratories
XV Special environments (practical work)
. * Vacuums, high altitudes, space, etc.
. * Alternative (not recommended) to treating as special forms of enquiry. See AY7 WB and the notes there. If this option is taken, proceed as follows:
. * Add to AY3 X letters W/Y following AY7: eg Vacuums AY3 XX.
Y *Special to a context*
. * For example of use, see BT3 YL Wind tunnels.
AY4 Instruments, instrumentation
. * As distinct from equipment and plant generally, instruments form a class of devices which perform a particular variety of operations in investigation: measuring, counting, monitoring, scanning, indicating, imaging.
. * Add to AY4 3 letters BD/M and R/T following AY3 U with the additions indicated; eg
AY4 37 . Unwanted effects
37I . . Aberration
. . *Special to instrumentation*

[Practical scientific work AY36]
[Agents]
[Equipment & materials together AY3B]
[Equipment & plant AY3U]
[Instruments AY4]
[Unwanted effects AY437]
. [Special to instrumentation]

AY4 38K . . Interference
38L . . . Self-interference, instrumental error
. . . . * Modification by the probe, etc. of the property under examination.
. . . . * See also error analysis (evaluation) AY7 DA
38M . . . Environmental interference (instrumentation)
38N . . . . *By form of interfering energy*
. . . . . * Add to AY4 38N letters A/Q following B; e.g.
38N J . . . . . Magnetic interference
38N KJ . . . . . Ionizing radiation
38O . . Drift (instrumentation)
38P . . Faults (instrumentation)
. . . * Avoidable engineering defects.
Operations on instruments
3BF . Installation & use
3BR . Performance criteria control
3BR W . . Allowances
3BS . . Clearances
3BT . . Tolerances
3BU . . . Limits & fits (instrumentation)
3BV . . . . Specific fits
. . . . . * Arrange A/Z; eg sliding fits AY4 4BV S.
3BW . . Calibrating
3BW 7A . . . Testing
3BW 7CO . . . . Standard test objects (calibration)
3BW S . . . Objective calibration
3C . Handling
Processes in instruments
3G . Deterioration in instruments
Properties of instruments
3JC . Characteristics
3JE . Reliability
3JL . Gain
3JM . Attenuation
3JN . Signal-to-noise ratio
3R *Materials of instrumentation*
Parts
. * Add to AY4 4 letters A/Y following AY3 N, with the additions shown; eg
4B . Surfaces
4H . Housings
4P . Power input sources (instrumentation)
5 Components in instrumentation

Practical scientific work

[Operations & agents]
[Practical scientific work AY36]
[Agents]
[Equipment & materials together AY3B]
[Equipment & plant AY3U]
[Instruments AY4]
[Components in instrumentation AY45]

* This class takes literature on components of instruments (which are used in a variety of types of instruments) and discusses the component as such rather than the whole instrument in which the component is used.
* Documents sometimes consider a component only as it functions within (and is used to define) a particular type of instrument. Such documents are classed under AY5 Types of instruments, in the array ((By possession of a particular component)); eg AY5 HS Fluidic instruments (meaning instruments with fluidic components).
* The retroactive rule for compounding may occasionally be broken since one component may have another component of its own, filing later than itself; eg AY4 SL4 ST Input tranducers - Receivers.
* Add to AY4 5 letters UV/V following AY3; eg

AY4 5V Intelligent instruments
6T Information processing instruments (components)
7 . Computers (instrument components)
8 Control instruments
. * The following is an extract from a larger schedule in U/V Technology, from which further details are obtainable. Notation is provisional.
82M . Mathematical techniques
837 . Unwanted effects
. *Properties*
83J S . . Stability (control systems)
83J T . . Time response (control systems)
. *Components*
847 . . Computers
84L . . Transducers
84L 4N . . . Indicators

[Practical scientific work AY36]
[Instruments AY4]
[Components in instrumentation AY45]
[Control instruments AY48]
. [Components]
. . [Transducers AY484L]
. . . [Indicators AY484L4N]

AY4 84T SW . . Windows
84V . . Regulators (instrumentation)
84W . . Filters
84X . . Actuators
84Y C . . Compensators
84Y E . . Energizing systems
. *Types of control systems*
. . *By form of working energy*
. . . * Add to AY4 86 letters A/J following AY4: eg
86B . . . Electronic
. . *By input-output structure*
8I . . . Open loop control systems
8IR . . . . Remote control
8J . . . Closed loop control systems, feedback control systems
8K . . . . Servomechanisms
. . *By general systems concepts*
. . . * Add to AY4 9 letters K/P following AY; eg
9M . . . Non-linear control systems
Types of instrument components by energy system
. * Add to AY4 letters A/T following U in U/V Technology.
. * Notation for the examples below is provisional until publication of Class U/V Technology.
AC . Electrical & electronic components
AD . . Electrical converters
AE . . . Rectifiers
AF . . Transformers
AG . . Capacitors
AH . . Resistors
AJ . . Switchgear
AL . . Relays
AM . . Conductors
AN . . Insulators
AP . . Circuits (instruments)
. . . * For circuits special to a given context, see context; eg for integrated circuits, see AY4 BB.
AV . . Excitation systems (instrument components)
. . . * Applying an electric signal to drive a device (eg an amplifier).
AW . . . Transmitters
AX . . . . Waveguides
. . . . . * Completely shielded transmission lines.
B . . Electronic instrument components
BB . . . Integrated circuits
BD . . . Electron tubes
C . . . Semiconductors
CP . . . Superconductors

[Practical scientific work AY36]
[Instruments AY4]
[Components in instrumentation AY45]
[Types of instrument components by energy system]
[Electrical & electronic components AY4AC]
. [Electronic instrument components AY4B]
. . [Superconductors AY4CP]

AY4 CS . . Stimulated emission devices (components)
CV . . . Masers (as components)
D . . . Lasers (as components)
DS . . Resonators
DT Telecommunication components (instruments)
DU . Telemetering components (instruments)
DW Electromagnetic components
E Magnetic components
EL . Magnetostrictive components
EO . Nuclear magnetic resonance (instruments)
F Optical components
. *Properties*
F3Q . . Scales (optical components)
. . . * Range of brightness or density for reproduction.
. *Types of optical components*
. *By energy form*
FH . . Photovoltaic components
FJ . . Photoconductive components
FK . . Laser excited optical components
FL . . Optoelectronic components
FN . . . Photomultiplier tubes
FP . . Photonic devices
. . . * Uses light rather than electrons to transmit signals in computers, etc.
FR . . Opto-acoustic components
. *By function*
. . * A detailed schedule for these components is given under Optics at BRL 4U/X.
. . * Add to AY4 F letters U/X following BRL; eg
FUF . . Filters (optical components)
FV . . Lenses
FWB . . Prisms
. . . * Refracting media which deviate and/or disperse light.
FWG . . Gratings
FWM . . Mirrors
. . . * Reflect light without significant dispersion.
G Thermal engineering components (instruments)
GH . Thermoelectric components
GJ . Thermomagnetic components
GM . Heat exchangers (instruments)
GN . Thermal insulation (instruments)

[Practical scientific work AY36]
[Equipment & plant AY3U]
[Instruments AY4]
[Components in instrumentation AY45]
[Types of instrument components by energy system]
. [Thermal engineering components instruments AY4G]
. . [Thermal insulation instruments AY4GN]

AY4 GP . . Heat energy sources (instrument components)
GR . . Refrigeration components (instrumentation)
H . Mechanical engineering components (instruments)
HL . . Linkage mechanisms (instruments)
HM . . Mechanical transmission (instruments)
HN . . . Couplings (instrument components)
HP . . . Bearings
HQ . . Damping devices
HS . . Fluidic components, fluidics
. . . * Fluid flow simulates electron flow in conductors; the interactions of streams of fluid is used to control instruments, etc.
HT . . Piezoelectric components
HU . . Acoustoelectric components
HV . . Photoelastic components
HW . . Acoustic components
HX . . . Ultrasonic components
J . Chemical components (instruments)
JL . . Electrolytic components
Types of components by internal function
. * Acting on other parts or features of the instrument.
K . Switching devices (instrument components), conversion devices, converters
KN . . Analogue-digital converters, digitizers
KP . . Digital-analogue converters
L . . Transducers, converters (transducers)
. . . * Converts one physical parameter (eg sound) into another (usually electrical or optical signals).
L4A V . . . Excitation
. . . *Types of transducers*
LN . . . . Non-self-exciting transducers
LP . . . . Self exciting transducers
LR . . . . Digital transducers
. . . . *By form of energy input*
. . . . . * Add to AY4 M letters A/J following AY4; eg
MA . . . . . Electrical transducers
MAG . . . . . . Capacitive transducers, capacitance transducers
MAH . . . . . . Resistive transducers
MAI . . . . . . Reductance transducers

[Practical scientific work AY36]
[Types of components by internal function]
[Switching devices instrument components AY4K]
. . [Types of transducers]
. . . [By form of energy input]
. . . . [Electrical transducers AY4MA]
. . . . . [Reductance transducers AY4MAI]

AY4 MCW . . . . . Electromagnetic transducers
MCX . . . . . . Inductive transducers, inductance
. . . . . . transducers
MD . . . . Magnetic transducers
MDL . . . . . Magnetostrictive transducers
MDO . . . . . Nuclear magnetic resonance
. . . . . transducers
ME . . . . Laser excited transducers
MF . . . . Optical transducers
MFH . . . . . Photovoltaic transducers
MFJ . . . . . Photoconductive transducers
MFL . . . . . Optoelectronic transducers
MFQ . . . . . Opto-acoustic transducers, acousto-
. . . . . optical transducers
MG . . . . Thermal transducers
MGM A . . . . . Thermoelectric transducers
MGM D . . . . . Thermomagnetic transducers
MH . . . . Mechanical transducers
MHR . . . . . Electromechanical transducers
MHS . . . . . Fluidic transducers
MHT . . . . . Piezoelectric transducers
MHU . . . . . Acousto-electric transducers
MHV . . . . . Photoelastic transducers
MHW . . . . . Acoustic transducers
MHX . . . . . . Ultrasonic transducers
MJ . . . . Chemical transducers
MJL . . . . . Electrolytic transducers
N Indicators (instrument components), pointers (instruments)
NN . Mechanical indicators
. . * Needles, pointers, etc.
NR . Radiation indicators
Q . Optical indicators (instrument components)
QV . Other indication devices
. . * Arrange A/Z.
R Recording devices (instrument components), recorders (instrument components)
RQ . Automatic recorders
RR . Coded recording devices
RS . Graphic recording (instrument
. components)
S Input devices
S4A V . Excitation systems
S4A W . . Transmitters
. Transducers
SL . . Input transducers
. . . *Parts*
SL4 ST . . . . Receivers
. . . *Types of input transducers*
. . . . *By location of environment*

[Practical scientific work AY36]
[Components in instrumentation AY45]
[Types of components by internal function]
. [Input devices AY4S]
. . [Transducers]
. . . . [Types of input transducers]
. . . . . [By location of environment]

AY4 SLR . . . . . . Airborne transducers
SLS . . . . . . Space vehicle borne transducers
SLT . . . . . . Underwater transducers
SLU . . . . Underseas transducers
SLV . . . . . . Subterranean transducers
SS . . Sensors, detectors, probes
SS3 V . . . Intelligent sensors
SSV . . . Image sensors
SSW . . . Tactile sensors
ST . . Receivers
. . . * Parts in contact with the phenomena
. . . measured.
. . . * For remote indication, see AY7 4M.
T . Output devices
. . Transmitters
T4A W . . . Signals transmission (technical
. . . operations)
TR . . Recorders
TRT . . . Chart recorders
TRU . . . X/Y plotters
TRV . . . Oscillographs
TRW . . . Digital recorders
TRW M . . . . Magnetic tape recorders, tape
. . . . recorders
TRX . . . Event recorders
TRY . . Viewing devices, visualization devices,
. . display devices
. . . * Visualization illuminates only; it doesn't
. . . necessarily form an image. For Imaging, see
. . . AY7 I.
. . . *Properties*
TSC . . . . Contrast
TSD . . . Dials
TSS . . . Scales
TSW . . . Windows (instrument controls)
TTC . . . Cathode ray tube displays
TTE . . . Electroluminescent display
TTL . . . Liquid crystal displays
TU . . Intensifiers, amplifiers (intensifiers)
V *Types of instrument components by special function*
. * 4V/4Y are reserved for these. They are used, eg, at
. AY4 84V Regulators.
. * If a special component is used in another context,
. details may be taken from the relevant classes; eg
. Optical instruments -4U (from BRL 4).

[Practical scientific work AY36]
[Agents]
. [Equipment & materials together AY3B]
. . [Equipment & plant AY3U]
. . . [Instruments AY4]
. . . . [Components in instrumentation AY45]
. . . . . [Types of instrument components by special function AY4V]

. . . . *Types of instruments*
. . . . . * Add to AY5 3 letters J/V following AY3 so far as applicable; eg
AY5 3V . . . . . Intelligent instruments
. . . . . *By possession of a particular component*
. . . . . . * Add to AY5 numbers & letters 5/U following AY4; eg
D . . . . . . Laser-driven instruments
HS . . . . . . Fluidic instruments, instruments with fluidic components
V . . . . . *Types of instruments special to a context*
. . . . . *Types of instruments by investigative function*
. . . . . . * The general arrangement in BC2 is to locate an instrument under the function it serves, wherever that may be in the general classification; e.g. measuring instruments AY7 64; flowmeters BS7 64; Rate flowmeters BSD C76 4.

Operations in scientific investigation
. * The central operation in much practical scientific investigation is a combination of observation and measurement. This is supplemented by numerous contributory activities such as sensing, detection, indication, recording, etc. In many cases it may be unhelpful to try to draw fine distinctions between these activities, so omnibus headings are provided for when the literature is too broad in coverage to justify making such distinctions.
. * Compounding of operations and agents should be done with caution. Often, one element in a potential compound is implicit in the other; e.g. measurement often implies the initial operation of observation; testing and evaluation usually implies measurement. Telemetry reflects the dual operation of indication and measurement together.

[Science in general AY2]
[Operations & agents]
[Practical scientific work AY36]
[Operations in scientific investigation]

AY6 2 Investigative techniques (general)
. *Operations serving all techniques & objectives*
. . * For analysis, see AY7 2N.
3 . . Data handling, data processing
4B . . . Automated techniques
4C . . . Computer techniques
4D . . . . Programs
4E . . . . Digital techniques
4KN . . . . . Analogue-digital conversion
4R . . . Data recording & display
. *Operations by scale*
6 . . High-speed techniques
. . . * Taking a microsecond or less.
7 . . Microtechniques (general)
. . . Instruments
74V . . . . Micromachines
8 . . . Nanotechniques
. . . . * On scales approximating the size of atoms and molecules.
8I . . Semi-microtechniques (general)
8M . . Macrotechniques (general)
. *Operations by energy form*
9 . . Physical methods in investigation
. . . * Add to AY6 numbers & letters 9,A/W following B6. A selection of major concepts is given here to show scope. More detail will be available when the technology schedules are published.
B . . . Mechanical techniques
E . . . Sonic vibration techniques
GH . . . Acoustic techniques
GP . . . Thermal techniques
HY . . . Electromagnetic & electronic techniques
IB . . . . Electronic techniques
J . . . . Magnetic techniques
K . . . . Radiation techniques electro-magnetic
KFT . . . . . Scatter techniques
KQ . . . . Microwave & optical techniques together
KQM . . . . . Stimulated emission techniques
KR . . . . . . Maser techniques
KS . . . . . . Laser techniques
KTV . . . . . Non-linear optical techniques
KTX . . . . Electrical engineering techniques
. . . . . * Add details from U Technology (when published).
KU . . . . Telecommunications techniques
. . . . . * Add details from U Technology (when published).

[Operations & agents]
[Practical scientific work AY36]
[Investigative techniques general AY62]
[Operations by energy form]
. [Physical methods in investigation AY69]
. . [Electromagnetic & electronic techniques AY6HY]
. . . [Telecommunications techniques AY6KU]

AY6 KV . . . . Microwave techniques
. . . . . *** Add to AY6 KV letters U/W following BK**
. . . . . **if applicable; eg AY6 KVV B Superhigh**
. . . . . **frequency techniques.**
KW . . . . Radio & television techniques
KY . . . . Radar techniques
L . . . Optical techniques
LFP . . . . Polarization optical techniques
LFR . . . . Interferometry
LM . . . . Colour optical techniques
. . . . *Forms of light*
LPF . . . . . Coherent light techniques
. . . . *By wavelength*
LU . . . . . Infrared techniques
LW . . . . . Ultraviolet techniques
LWY . . . Radiological techniques
LX . . . . X-ray techniques
M . . . Particulate radiation techniques,
. . . particle physics techniques
MGB . . . . Particle beam techniques
. . . . *By particle*
. . . . . *** Add to AY6 letters N/Q following B; eg**
NGO . . . . . Photon techniques, photonics
NP . . . . . Electron techniques
O . . . . . Nuclear reaction techniques
OFK . . . . . . Radioactivity techniques
. . . . *** For tracer techniques, see AY7 P.**
QU . . . . . Ion techniques
QX . . Vacuum techniques
. . . *** For vacuums as a form of matter, see BRY.**
R . . Bulk matter techniques
. . . *** For sonic techniques and thermal techniques,**
. . . **see AY6 E.**
. . . *** Add to AY6 letters R/W following B; eg**
RNR . . . Phase transition techniques
TCW . . . Gas dynamics techniques
UCH . . . Hydrostatic techniques
UCW . . . Hydrodynamic techniques

X . . . Chemical techniques
. . . . *** Add to AY6 X letters A/S following C**
. . . . **(notation is provisional); eg**
XB . . . . Chemical analysis (techniques)
XCE . . . . Electrochemical techniques
XCE L . . . . . Electrolytic techniques
XX . . . Biological techniques
Y . *Operations special to a context*
. . *** Eg AY7 76Y C Conversion (of measurement**
. . **systems).**
By action on the phenomena investigated

[Science in general AY2]
[Operations & agents]
[Practical scientific work AY36]
[Investigative techniques general AY62]
[By action on the phenomena investigated]

*** Many of the physical methods of investigation above (at AY6) serve several operations; eg X-ray techniques may serve detection, indication, monitoring, imaging etc. The classes below take that literature which deals primarily with the specific operations themselves, regardless of technique used.**
AY7 2 Control techniques
2B . Hysteresis (control techniques)
2D . Damping (control techniques)
2E . . Isolation damping
2N Analysis (of phenomena investigated)
. *** For example, analysis of sound.**
2P . Qualitative analysis
2Q . Quantitative analysis
3 Production techniques
. *** Production of the phenomena**
. **investigated; e.g. light sources in optical**
. **investigations.**
3B . Sampling (production techniques)
3C . . Preparation of sample
3D . Generation techniques,
. reproduction techniques
3F . . Computer generation (production
. . techniques)
3H . . Synthesis (of phenomena
. . investigated)
3K . *By source. medium*
. . *** For example, Sound - Blowing**
. . **BRG H73 LB.**
3L . *Special to a context*
. . *** For example, separation in particle**
. . **production BM7 3P.**
4 Observing
4D . Direct sensory perception of
. phenomena
4G Detecting & indicating (together)
. *** See also detecting & counting (together)**
. **AY7 5**
4J . Detecting, detection, sensing
4L . Indicating, display
. . *** See also Visualizing & imaging AY7 H**
4M . . Remote indication
4N . . . Telemetry
. . . . *** Presenting results at a distance.**
4P . . . Echo probing
4Q . . . . Radar (indicating), radio echo
. . . . sounding
4T Recording
4TR . Interpretation
4U . Graphic representation

Investigative techniques (general)

[Science in general AY2]
[Operations & agents]
[Practical scientific work AY36]
[Investigative techniques general AY62]
[By action on the phenomena investigated]
[Recording AY74T]
. [Graphic representation AY74U]

AY7 4V . Scanning
. . * Giving information on conditions at
. . various positions in space, examined
. . cyclically.
. . * See also Monitoring AY7 C
4W . Identifying
5 Counting, counting & detecting (together), calculation (counting)
. * Include here works on counting
. techniques alone.
. * See BM7 5 for a detailed schedule
. relating to particles.
. * See also Visualization & counting
. (together) AY7 GY.
6 Measurement, mensuration, metrology
62M . Mathematical methods in
. measurement
638 E . Errors in measurement
. Instrumentation
64 . . Measuring instruments, gauges,
. . meters
. . . * For photogrammetry, see AY7 KV.
643 BKE . . . Standards
69 . Principles of measurement
. . * For measurement of things by a
. . multiplicity of criteria representing
. . desiderata in some context, see
. . Standards AY7 E.
. . * See also Performance criteria AY7 EP
69B . . Summation methods, integral
. . measurement
69D . . . Micromeasurement
6A . . Units of measurements, systems
. . of units, standard units
. . . * For SI units, see Systems of
. . . measurement AY7 7M.
6AD . . . Dimensions of units
6AL . . . Absolute units
6AR . . . Relative units, practical units
6C . . Constants
. . . * For tables of constants, see AY2 3NL
6C6 YC . . . Conversion
. *Techniques of measurement*
6D . . Direct measurement
6F . . Differential methods
. . (measurement), small
. . differences method
. . (measurement)
6H . . Maxima & minima (measurement)

[Operations & agents]
[Practical scientific work AY36]
[Investigative techniques general AY62]
[By action on the phenomena investigated]
[Measurement AY76]
[Techniques of measurement]
. [Maxima & minima measurement AY76H]

AY7 6J . Performance characteristics in
. measurement
. . * Interaction of instrument & measurand.
6K . . Precision
6L . . Signal-to-noise ratio
6M . . Dynamic response
6N . . Drift (measurement)
6O . . Hysteresis (measurement
. . techniques)
6P . Presentation characteristics
. (measurements)
. . * For telemetry, see AY7 4N.
6Q . . Digitalization
6R . . Amplification
6S . . . Magnification
. . . . * Enlargement without increasing
. . . . power of signal. For microscopy, see
. . . . AY7 J.
6T . *Special to a subject*
. . * For example, under pyrometry
. . BRG X76.
7 Systems of measurement
. *Operations*
76Y C . . Conversion (measurements)
76Y C2 . . . Tables
7C . Weights & measures
. . * Originally consisted of four measures -
. . mass, volume, distance, area. Now
. . includes other parameters also (e.g.
. . temperature, luminosity).
7E . Metric system
7G . CGS system
. . * Centimetre, gram, second; now largely
. . replace by MKS and SI.
7J . MKS system
. . * Metre, kilogram, second
7K . MKSA system
. . * Amplification of, & now largely
. . replacing, MKS systeml includes
. . amperes, etc.
7M . Systeme Internationale d'Unitees,
. SI (measurement)
7P . Imperial system (measurement)
8 . Time variables measurement
84 . . Instrumentation
. . . *Components*
84V C . . . . Hands (clocks & watches)
84V E . . . . Winding mechanisms
84V G . . . . Movements (clocks & watches)
. . . *Types of time instrumentation*

Investigative techniques (general)

[Investigative techniques general AY62]
[By action on the phenomena investigated]
[Measurement AY76]
. [Systems of measurement AY77]
. . [Time variables measurement AY78]
. . . [Instrumentation AY784]
. . . . [Types of time instrumentation]

AY7 85V K . . . . . Clocks & watches
85V L . . . . . . Electronic clocks
85V M . . . . . . Atomic clocks
85V P . . . . . Chronometers
85V R . . . . . Timing devices
85V S . . . . . . Start devices
85V T . . . . . . Stop devices
85V V . . . . Stop clocks & watches
88E . . . Time standards
8B . . . Frequency measurement
8D . . . Calendars
. . . . * See Class D Astronomy & Earth sciences.
8J . . Space variables measurement systems
8L . . Scales
. . . * For example, seismic scales in Class D.
8Q . . *Systems of measurement special to a context*
. . . * For example, BRL L78 R Visual photometry
9 . . Other variables measured
. . . * The preferred arrangement is to subordinate the measurement of a specific thing to the latter. This location is an alternative (not recommended) for libraries wishing to keep all works on measurement together.
. . . * If this option is taken, proceed as follows: add to AY7 9 letters A/Y following B; eg Density measurement AY7 9CL.
A Testing & evaluation
. * Measuring something against an agreed standard which conforms to some set of desired criteria, which together define an ideal model.
. * For technical testing, see Technology UE
A3U . Equipment, laboratories, facilities
A3Y . . Test chambers
AB . Inspection, examination
AE . Testing, trials, tests
AEF . . Preliminary tests
AEH . . Working tests, operational tests, in-house tests
AEM . . Re-testing, post-modification tests
B . . Technical testing
. . . * Add letters from UE Technical testing (technology) (not yet published).
C . . Monitoring
. . . * Testing over time; periodic or continuous determination and/or verification of a variable.
D . Evaluation, assessment, appraisal, estimating (evaluation), valuation
DA . . Error analysis
. . . * For errors as a source of hazards, see AY3 8E.

[Operations & agents]
[Practical scientific work AY36]
[Investigative techniques general AY62]
[By action on the phenomena investigated]
[Testing & evaluation AY7A]
. [Evaluation AY7D]
. . [Error analysis AY7DA]

AY7 DC . . . Accuracy (errors)
DD . . . Correction of error
. . . . * See also Feedback control AYP J
. . . *Types of errors*
DDE . . . . Mean error
. . . . *By source*
DDG . . . . . Observer error
DDI . . . . . Instrument error
DDK . . . . . Transmission media error
DDM . . . . . Object of measurement error
. . *Types of evaluations*
DE . . . Expert assessment
DF . . . Current assessment
DG . . . Retrospective assessment
DH . . . Prospective assessment
DK . . . Undervaluation
DO . . . Overvaluation
. . *Products of evaluation*
E . . . Standards (evaluation), desiderata (standards)
. . . . * For social agreements on and administration of standards generally, see AY1 29U.
EH . . . . Quantitative standards, dimensional standards
. . . . . * See also Systems of measurement AY7 7
EL . . . . Qualitative standards, performance standards
. . . . . * For quality assurance and quality control, see Technology UF.
. . *Subjects of evaluation*
. . . * The general arrangement is to subordinate evaluation to the phenomenon evaluated. Below, a number of very generalized attributes of action, too diffuse to warrant citing first, is given.
EP . . . Performance evaluation, performance criteria
. . . . * See also Performance criteria control AY3 BR
EQ . . . . Efficiency (evaluation of)
ER . . . . Reliability evaluation
ES . . . Problem evaluation
ET . . . . Risk evaluation
EV . . . Project evaluation
F Modelling, simulation, models
. * See also Statistical models AXQ R
FAE . Testing
FG . Solving (models)

Investigative techniques (general)

[Science in general AY2]
[Operations & agents]
[Practical scientific work AY36]
[Investigative techniques general AY62]
[By action on the phenomena investigated]
[Modelling AY7F]
. [Solving models AY7FG]

AY7 FH . Predicting (models)
. *Types of models*
FM . . Mathematical & statistical models
. . . * Add to AY7 FM letters B/Y following AM4 Q;
. . . * Add to AY7 FN letters A/E following AM4 R; eg AY7 FMT Abstract models.
FX . . . Probabilistic models
FYC . . Causal models
FYE . . Natural models
FYG . . Scale models
FYI . . Iconic models
FYK . . Structural models
FYL . . . Graph models
FYN . . Static models
FYP . . Dynamic models, process models
FYS . . Physical models
G Prediction, forecasting
GY Visualization & counting (together)
. * See BM7 H Track visualization (particles) for a special application of this class.
. * For counting (general), see AY7 5.
H Visualizing & imaging (together)
. Instrument components
H4 . . Display devices (visualizing & imaging)
HV . Visualizing (techniques)
. . * Illuminates only; does not necessarily form image.
I . Imaging (techniques)
I6M GB . . Particle beam techniques
. . . * Use AY7 IM.
. . *Operations*
ICN . . . Optical transfer function
. . . . * Mathematical representation of the effects on imaging of lenses or other components.
ICP . . . . Phase transfer function
ID . . . Beam handling (general)
. . . . * Add to AY7 I letters DF/F following BM7 IM; eg
IDF . . . . Beam defection
IE . . . . Focusing
IEG . . . . . Beam trapping
IF . . . Resolution(
IFH . . . Modulation
IFS . . . Scanning
IFT . . . Projection

[Practical scientific work AY36]
[Investigative techniques general AY62]
[By action on the phenomena investigated]
[Visualizing & imaging together AY7H]
[Imaging techniques AY7I]
[Operations]
. [Projection AY7IFT]

AY7 IFV . . Collimation
Properties
. *Beam properties*
. . * Add to AY7 IG letters B/V following BM7 IMG; eg
IGB . . Beam profile
Types of imaging by radiation form or particle
. * Add to AY7 I letters H/Q following B; eg
ILU . Infrared imaging, thermal imaging, thermography
IM . Particle beam techniques
. . * Focusing of particle beams to form images, analogously to the formation of light beams in optics.
. . * The detailed schedule for this is given under particle physics BM. Details may be taken from there if needed.
. . * Add to AY7 IM numbers & letters 4/G following BM7 IM;
. . * Add to AY7 I letters N/Q following B; eg
. . *Instrument components*
IM4 FV . . . Lenses
IM4 FVV . . . . Electrostatic lenses
IM4 FVW . . . . Magnetic lenses
. . *Properties*
IMG B . . . Beam profile
. . *Types of particles*
INP . . . Electron beams
INV . . . Proton beams
. . . . * For proton resonance imaging, see AY7 IO.
IO . . . Nuclear magnetic resonance imaging
IQU . . . Ion beams, ion optics
Types of imaging by other characteristics
IST . Stereographic projection, stereograms
IX Magnification techniques, amplification techniques (images)
J . Microscopy
. . Instrumentation
. . . Components
J4F UF . . . . Filters
J4F V . . . . Lenses
J4W B . . . . Prisms
. . . *Types*

[Investigative techniques general AY62]
[Visualizing & imaging together AY7H]
[Imaging techniques AY7I]
[Magnification techniques AY7IX]
[Microscopy AY7J]
[Instrumentation]
. [Types]

AY7 J5 . . Simple & compound microscopes
Types of microscopy by illumination
JED . Dark field microscopy, dark ground microscopy
JEF . Field emission microscopy
JEH . Field ionization microscopy
JEJ . Fluorescent microscopy
JEL . Phase contrast microscopy
By direction
JEN . Vertical field microscopy
JES . Stereoscopic microscopy
By property
. * Add to AY7 JF letters B/Y following BF; eg
JFN . Reflecting microscopy
JFP . Polarizing microscopy
JFQ . Diffraction microscopy
JFR . Interference microscopy
JFR 7K . . Holographic microscopy, holomicrography
By wavelength
. * Add to AY7 J letters G/Q following B, with the modifications indicated:
JGH . Acoustic microscopy
. . * Using acoustic waves at microwave frequencies.
JGL . . Ultrasonic absorption microscopy
JL . Optical microscopy
. . * Visible light usually assumed. Includes documents dealing with i-r and u-v light as well.
JLU . . Infrared microscopy
JLW . . Ultraviolet microscopy
JLW N . . . Flying spot microscopy
JLX . X-ray microscopy
. . * See also Electron microscopy AY7 JQ
JLX G . . X-ray projection microscopy
JLX L . . X-ray absorption microscopy
JLX Q . . X-ray diffraction microscopy
JM . Particle microscopy
. . Instrumentation
JM5 V . . . Ultramicroscopes
JNP . . Electron microscopy
. . . * Use AY7 JP.
. . . * This interrupts the division by NQ/Q, which is resumed at AY7 JS.
JP . . Electron microscopy
. . . *By characteristics of specimen*
JPM . . . . Specimen cooling electron microscopy

[Investigative techniques general AY62]
[Microscopy AY7J]
[By wavelength]
[Particle microscopy AY7JM]
. [Electron microscopy AY7JP]
. . [By characteristics of specimen]
. . . [Specimen cooling electron microscopy AY7JPM]

AY7 JPN . . . Specimen heating electron microscopy
. . *By component*
JPP . . . Mirror electron microscopy
JPR . . . Replica electron microscopy
. . *By radiation property*
. . . * Add to AY7 JQ letters C/U following BF, with modifications; eg
JQC . . . Transmission electron microscopy
JQG . . . Emission electron microscopy
. . . . * See also Field ion electron microscopy AY7 JXP
JQH . . . . Field emission electron microscopy
JQN . . . Reflection electron microscopy
JQQ . . . Diffraction electron microscopy
JQQ T . . . . Transmission diffraction electron microscopy
JQQ V . . . . Selected diffraction electron microscopy
JQQ W . . . . . Shadow diffraction electron microscopy
. . *By special method*
JRG . . . Gas reaction electron microscopy
. . . . * See also Field ion electron microscopy AY7 JSU P
JRN . . . Thin film electron microscopy
JRS . . . Shadow electron microscopy
JRT . . . Scanning electron microscopy
JRT T . . . . Scanning-transmission electron microscopy
JRT V . . . High voltage electron microscopy
. . *By focusing system*
JRW . . . Electrostatic electron microscopy
JRX . . . Electromagnetic microscopy
JS *Other particle microscopies*
. * Division by BM/BQ is resumed here after its interruption at AY7 JP.
. * Add to AY7 JS letters NPR/Q following B; eg
JSO . Nuclear magnetic resonance microscopy
. . * For nuclear magnetic resonance imaging, see AY7 IO.
JSU . Ion microscopy
JSU N . . Field ion microscopy
JSU P . . . Electron (field ion) microscopy
JSU R . . . Field emission ion microscopy
JSU S . . . . Atom probe field ion electron microscopy

Investigative techniques (general)

[Investigative techniques general AY62]
[Magnification techniques AY7IX]
[Microscopy AY7J]
. . . [Ion microscopy AY7JSU]
. . . . [Field ion microscopy AY7JSUN]
. . . . . [Field emission ion microscopy AY7JSUR]
. . . . . . [Atom probe field ion electron microscopy AY7JSUS]

AY7 K Holography (techniques), wavefront reconstruction imaging
. * Imaging technique without using camera or lenses. Records and reconstructs the wavefront emanating from an object illuminated by a laser beam.
K55 . Instrument components
K5F . . Optical elements
K5F V . . . Holographic gratings, holographic plates
K5X . . Holograms, interferometers (hologram)
. Recording
K74 T . . Recording (holographic)
. *Processes*
. . * Add to AY7 KF letters B/Y following BF; eg
KFQ . . Diffraction
. . Interferometry
KFR . . . Interferometry (holographic)
. *Product elements*
KHR . . Real image
KHS . . Virtual image
. *Types*
KK . . Computer generated holography
. *Types of holography by wavelength*
KLH . . Acoustic holography
KLJ . . . Holographic multiplexing
KLL . . Ultrasonic holography
KLQ . . Microwave holography
KLX . . Optical holography
KM . . Particle field holography
. . . * Add to AY7 K letters M/Q following B if applicable.
. *Types of holography by technique*
KRF . . In-line Fraunshofer holography
KRH . . High-speed holography
KT Photographic techniques
KV . Photogrammetry
. . * Use of photographic records for mapping, measurement of dimensions, etc (eg by aerial surveying).
L Radiography
. * Use of shortwave ionizing radiation to produce images.
L . *By radiation source*
. . * Add to AY7 L letters L/Q following B; eg
LLX . . X-ray radiography
LLX T . . Gamma ray radiography
LNV . . Proton radiography
LNW . . Neutron radiography

[Investigative techniques general AY62]
[Visualizing & imaging together AY7H]
[Imaging techniques AY7I]
. [Magnification techniques AY7IX]
. . [Radiography AY7L]
. . . [By radiation source AY7L]
. . . . [Neutron radiography AY7LNW]

AY7 LR . . . Autoradiography
LRS . . . . Shadowradiography
LRT . . . . Colour radiography
LRX . . . Flash radiography
LS . . . Microradiography
LSQ . . . Macroradiography
LSS . . . Stereoradiography
LV . . Tomography
. . . * Image of a selected plane in a material, etc. The material is examined by rotating a detector and the source of radiation in such a way as to blurr points outside the plane and thereby highlight the latter.
LV6 4C . . . Computerized tomography
M Spectroscopy, spectrography, spectrum analysis
. * Terminology varies; spectrometry is often equated with spectroscopy and works which do this should be classed here, not at AY7 N.
. * See also Scatter techniques AY7 KFT
. Instrumentation
M4 . . Spectroscopes
M5W . . Spectrographs
. *Operations*
MBD . . Excitation
MBE . . . Activation analysis
. *Processes measured*
MBG . . Continuous spectrum
MBH . . Line spectrum
MBJ . . Band spectrum
MC . . Radiation phenomena (spectroscopy)
. . . * See also types of spectroscopy specified by these (at AY7 MF).
. . . * Add to AY7 MC letters B/U following BF; eg
MCG . . . Emission
MCL . . . Absorption
MCP . . . Polarization
MCR . . . Interference
MCT . . . Scattering
. *By exciting agent*
MDC . . Arc spectroscopy
MDE . . Spark spectroscopy
MDF . . Flame spectroscopy
MDL . . Fluorescence spectroscopy
. *By radiation phenomenon*
. . * Add to AY7 MF letters B/Y following BF; eg
MFF . . Coherence spectroscopy
MFG . . Emission spectroscopy
MFL . . Absorption spectroscopy
MFN H . . Reflectance spectroscopy

Investigative techniques (general)

[Practical scientific work AY36]
[Investigative techniques general AY62]
[By action on the phenomena investigated]
[Visualizing & imaging together AY7H]
[Spectroscopy AY7M]
[By radiation phenomenon]
. [Reflectance spectroscopy AY7MFNH]

AY7 MFO	. Resonance spectroscopy
	. . * See also particular resonating energies or particles; eg Nuclear magnetic resonance
MFO Q	. . Quadrupole resonance spectroscopy
MFP	. Polarization
MFP	. . Rotational spectrum
	. . . * See also Chemical analysis C9M LCP R.
MFR	. Interference spectroscopy
MFS	. Collision spectroscopy
	By special wave properties
MGD	. Quantum beat spectroscopy
MGF	. Fourier transform spectroscopy
MGG	. Hadamand transformation spectroscopy
MGH	. Acoustic spectroscopy
MGJ	. Magnetic resonance spectroscopy
	. . * See also resonating particles; eg nuclear magnetic resonance AY7 MON.
	By wave/particle used
	. * For acoustic spectroscopy, see AY7 MGC
	. * Add to AY7 M letters K/Q following B; eg
MKM	. Radiofrequency spectroscopy
MKN	. . Nuclear magnetic resonance spectroscopy, NMR, nuclear magnetic spin resonance spectroscopy
MKN Q	. . . Nuclear quadrupole resonance spectroscopy
MKN R	. . . Pulse nuclear magnetic resonance
MKO	. . High frequency spectroscopy
	. Microwave & optical together
MKS	. . Laser spectroscopy
MKS R	. . . Raman laser spectroscopy
MKU	. Microwave spectroscopy
MKX	. . Electron spin resonance spectroscopy, ESR spectroscopy, electron spin magnetic resonance spectroscopy, electron paramagnetic resonance spectroscopy
MKX S	. . . Electron spin double resonance spectroscopy
MKY P	. . Atomic beam electron resonance (spectroscopy)
MKY R	. . Phosphorescence microwave double resonance, PMDR

[Investigative techniques general AY62]
[By action on the phenomena investigated]
[Visualizing & imaging together AY7H]
[Spectroscopy AY7M]
[By wave/particle used]
[Microwave spectroscopy AY7MKU]
. [Phosphorescence microwave double resonance AY7MKYR]

AY7 MKY S	. . Phosphorescence microwave photoexcitation spectroscopy, PMPS
ML	Optical spectroscopy
	. * If confined to visible light wavebend, see AY7 MLV.
	. *By property*
	. . * For coherence spectroscopy, see AY7 MFF.
MLF G	. . Emission spectroscopy
MLF O	. . Optical double resonance
MLF OH	. . . Microwave optical double resonance
MLP	. Light beating spectroscopy
MLQ	. Monochromatic light spectroscopy
MLR	. . Raman spectroscopy
	. . . * See also Raman spectrophotometry AY7 NSL NQ
MLS	. . Photoacoustic spectroscopy, optoacoustic spectroscopy
MLU	. Infrared spectroscopy, IR spectroscopy
MLU GF	. . Fourier transform infrared spectroscopy
MLV	. Visible light spectroscopy
MLW	. Ultraviolet spectroscopy, UV spectroscopy
MLW V	. . Vacuum ultraviolet spectroscopy
MLX	X-ray spectroscopy
	. * For Appearance potential spectroscopy, see Ion spectroscopy AY7 MUL.
MLX 7MN P	. Electron probe microanalysis
	. . * Examining X-ray spectra by electron beam.
MLY	Gamma ray spectroscopy
MLY M	. Mossbauer spectroscopy
MM	Particulate spectra techniques
MNG	. Photon spectroscopy
	. . * For photoemission spectroscopy (result of bombardment by photons) see Electron spectroscopy AY7 MNP.
MNH	. . Two photon spectroscopy
MNJ	. . Photon correlation spectroscopy
	. . . * See also Photo-electron spectroscopy AY7 MNP T
MNP	. Electron spectroscopy, electron emission spectroscopy
	. . * For electron spin resonance spectroscopy, see AY7 MKX.

Investigative techniques (general)

[Investigative techniques general AY62]
[By action on the phenomena investigated]
[Visualizing & imaging together AY7H]
[Spectroscopy AY7M]
[By wave/particle used]
. [Particulate spectra techniques AY7MM]
. . [Electron spectroscopy AY7MNP]

AY7 MNP Q . . . Electron energy loss spectroscopy
MNP S . . . Auger electron spectroscopy
MNP T . . . Photo-electron spectroscopy, photoemsssion spectroscopy
MNT . . Baryon resonance spectroscopy
MNW . . Neutron capture spectroscopy
MO . . Nuclear spectroscopy
. . . * For nuclear magnetic resonance spectroscopy, see radiofrequency spectroscopy AY7 MKN.
MP . . Atomic spectroscopy
. . . * See also Atom probe field ion microscopy AY7 JXS
MPL . . . Atomic absorption spectroscopy
MQ . . Molecular spectroscopy
MQM . . . Rotation spectroscopy
MQN . . . Vibration spectroscopy
MQP B . . . Band spectroscopy
MQP D . . . Time resolved spectroscopy
MQP F . . . Beam foil spectroscopy
MQU . . Ion spectroscopy, ion beam spectroscopy
MQU L . . . Appearance potential spectroscopy
MQU M . . . Secondary ion spectroscopy
MQU P . . . Ion microprobe analysis
MQU Q . . . Ion cyclotron resonance spectroscopy
By various physical constants
N . Spectrometry
. . * See note under Spectroscopy B7M. Prefer latter if spectrometry is equated with spectroscopy.
. . Instrumentation
N4 . . . Spectrometers
. . *By radiation/particle*
. . . * Add to AY7 N letters L/Q following B; eg
NLX . . . X-ray spectrometry
NLX T . . . Gamma ray spectrometry
NM . . . Particle spectrometry, mass spectrometry
NMN . . . . Cherenkov spectrometry, Cerenkov spectrometry
NNP . . . . Beta particle spectrometry, beta ray spectroscopy
NNW . . . . Neutron spectrometry
NO . . . Nuclear spectrometry
NOQ G . . . . Photoneutron spectrometry
NOX . . . . Alpha particle spectrometry
NS . Spectrophotometry
. . * Measures intensity of wavelength in optical spectra.

[Practical scientific work AY36]
[Investigative techniques general AY62]
. [By action on the phenomena investigated]
. . [Visualizing & imaging together AY7H]
. . . [Spectroscopy AY7M]
. . . . [By various physical constants]
. . . . . [Spectrophotometry AY7NS]

. . . . . . Instrumentation
AY7 NS4 . . . . Spectrophotometers
. . . . . . *By special wave properties*
. . . . * Add to AY7 NSF letters B/Y following BF; eg
NSF G . . . . Emission spectrophotometry
NSF L . . . . Absorption spectrophotometry, absorptiometry
NSH . . . . Differential spectrophotometry
. . . . . . *By radiation/particle*
NSL . . . . . . Optical spectrophotometry
. . . . Monochromatic
NSL NQ . . . . Raman spectrophotometry
NSL U . . . . Infrared spectrophotometry
NSL W . . . . Ultraviolet spectrophotometry
NSP . . . . . . Atoms
NSP FL . . . . Atomic absorption spectrophotometry
NX . . . . Mass spectrum analysis
O . . . . . Mass spectroscopy
. . . . . . * Measures atomic mass by separating beam of ions into components reflecting different mass/charge ratios.
OP . . . . . . Mass spectrographs
OQ . . . . . . Time of flight mass spectroscopy
OT . . . . . . Tunnelling spectroscopy
OU . . . . . . Quadrupole mass spectroscopy
OV . . . . . Mass spectrometry
. . . . . . Instrumentation
OV4 . . . . Mass spectrometers, velocity spectrometers
P . . Tracer techniques
P75 . . . Counting
P75 4 . . . . Counters
PE . . . Labelled compounds, tracers
PG . . . . Radioactive tracers, radioisotope tracers
PJ . . . . Tritium tracers, tritiaated tracers
Q . . *Techniques special to a subject*
S *Investigative procedures special to a subject*
. * Notation AY7 S/V is reserved for such operations; eg seismological techniques in Earth sciences; Conversion of optical frequencies BLF D7T; Acceleration of particles BM7 T

Practical scientific work

[Science in general AY2]
[Operations & agents]
[Practical scientific work AY36]
[Investigative procedures special to a subject AY7S]

Types of investigation by persons as agents
AY7 WD . Individual research
WG . Group research, team research, organized research
WJ . . Big science
WL Special environments (practical science)
. * These represent quasi-laboratory conditions affecting the conducting of experiments, etc.
. * For research into the environment, see Class D.
. * An alternative (not recommended) is to locate these with laboratories at AY3 XV.
. *Operations*
WN . . Control of environment in practical work
WP . Hypothetical environments
. . * For Aether, see BAC EH.
X . Vacuums (as research environments)
. . * Including pressures below one atmosphere.
. . * For vacuum as a state of matter, see BQX
. . Equipment
X3W . . . Auxiliary vacuum apparatus
X73 . . Production
X73 P . . . Evacuating power (vacuums)
X73 Q . . . . Residual gas (vacuums)
X73 R . . . . . Degasification, degassing (vacuums)
X73 S . . . Pumping (vacuums)
X73 S4 . . . . Vacuum pumps
. . Measurement
X76 4 . . . Vacuum meters, vacuum gauges
XC . High pressure techniques
XE . Subsurface (research environments)
XF . Submarine investigations, underwater investigations
XH . High altitude (research environments)
XJ . Research in space, zero-gravity research
. . * For research into space itself, see Class B.
XN Non-experimental research (general)
. * For theoretical investigation, see AY8 B.
XO . Observational research
. . * Usually implies biological context. For 'natural history', see note at AYB N
XR . Descriptive research
XU . Surveys (research)
. . * Applies mainly to social sciences.
YB . Scientific exploration, scientific expeditions
YD . . Specific expeditions
. . . * Usually within a particular scientific field; arrange by name A/Z.
YF . Field investigation
AY8 2 Experimental research (general), experiments, experimentation
. * Use this class only for general works which draw a clear distinction between experimental and practical work. See note at AY3 2. If in doubt, prefer AY3 6.

[Science in general AY2]
[Operations & agents]
[Practical scientific work AY36]
[Experimental research general AY82]

. * For non-experimental investigation, see AY7 XN.
. * An alternative (not recommended) is to locate this class at AY3 5 (where it will be collocated with general theory at AY3 4 if the alternative for that is also taken).
AY8 22X . Statistical methods
. . * Use AY8 2E.
2D . Design of experiments
2E . . Statistical methods
. . . * Add to AY8 2E letters A/Y following AX; eg
2ER . . Statistical design (of experiments)
2ES V . . Factor analysis (design of experiments)
. *Organization of experiments*
2QS . . Operations research in experimentation
. . . * For replication of research, see AY8 CJ.
. . . * Add to AY8 2 letters QS/R following AY2 Y; eg
2QU K . . . Optimization (experiments)
2QW Q . . . Queuing problems (experiments)
2T . Falsification (experiments)
Types of research by broad objective
. * In principle, types of enquiry defined by purpose should file after all other types. Because of its special status as a form of enquiry (see notes at AY3 2) theory files after this array in order to collocate it directly with the phenomena investigated (see AY8 A).
5 . Fundamental research, basic research, free research, pure research
6 . Applied research
6M . Research & development in science, R & D in science
. . * Research and development (R & D) is unequivocally technological and general works on it go at U. But in social policy relating to science it is often considered alongside purely scientific research; for this conjunction, see AY1 3AF.
. . * See also Planning & development in science AY2 9G
7 . Oriented research
7J . . Oriented free research
7L . . Discipline oriented research
7M . . Science missions (general), mission-oriented research
. . . * For missions reflecting specific fields, see latter; eg Space missions in Class D.
8 . Interdisciplinary research in science, multidisciplinary research
9C . Civil research
. . * Includes civil research which has military applications also.
9E . Exploratory research
9H . Adaptative research

Practical scientific work

[Science in general AY2]
[Operations & agents]
[Practical scientific work AY36]
[Types of research by broad objective]
. [Adaptative research AY89H]

AY8 B Scientific theory, theoretical models, theories (general)
. * For the role of theory in explanation, see Science as a disipline AY2 9XM. This class is for the operations involved in theorizing.
. * An alternative (not recommended) is to locate this class at AY3 4, preceding practical work.
. * For scientific laws, see AY2 9XS.
B2M . Mathematical models (general)
. . * If distinguished from scientific models generally.
. . * For particular mathematical techniques used in science, see AY2 M.
BF . Formulation of theories, construction of theories
BG . . Formulation of problem
BH . . Abstraction
BK . Hypothetico-deductive system
BL . . Induction (scientific method)
BM . . . Generalization (scientific method)
BQ . . Deduction (scientific method)
BT . . Classification, categorization
C . Hypotheses
CJ . . Testing hypotheses, validation (research), replication (research)
. . . * See also Experimentation AY8 2
CL . . Working hypotheses, heuristic hypotheses
CM . Concept formation & meaning
CN . . Meaning, interpretation, interpretative analysis
. . . * For explanation in science, see AY2 9XL.
. . . * For statistical analysis, see AY2 XS.
CS . . Causal analysis
CT . . Levels of analysis
. . . * Expressing functional relations between variables.
D . Theories & models particular to a subject
. . * For works under specific classes dealing with the theories per se; eg Particle physics - Grand unified theory BM8 KQ.
. . * An alternative (not recommended) is to locate this class preceding practical scientific work, at AY3 5.

[Science in general AY2]
[Operations & agents]
. [Practical scientific work AY36]
. . [Scientific theory AY8B]
. . . [Theories & models particular to a subject AY8D]

General processes/properties
. * Some of the concepts below, reflecting commonly occurring processes or properties (particularly the conditions affecting a phenomenon) duplicate provision made in the main schedules. Synthesis by the latter may not produce the most helpful order for these concepts and may entail long classmarks. In such cases, the concepts are better taken from here, alongside those truly common concepts which do not appear in the main schedules.
. * The classes -92D/9S below are for use only as qualifiers; they should not be used as substantive classes.

AY9 2D . Distribution
2E . . Incidence, occurrence
2F . Invariability, constancy
. . * For concepts of stability and instability as systemic features, see AYH U.
. . * For constants in measurement, see AY7 6C.
2H . Variation, change, variables
2J . . Rate of change, gradient
2L . . Decrease
2M . . Increase
2N . . Cyclical change
2S . . Development
2T . . . Formation, origin
2V . . . Growth
4 . *Conditions & parameters & environments of physical events*
. . * Environments of physical events; includes relations of influence, effect, etc.
4BC . . Degrees of freedom
4C . . Critical point, critical state (conditions)
4E . . Volume conditions
4F . . . Constant volume conditions
4G . . . Decreasing volume conditions
4H . . . Increasing volume conditions
4J . . Pressure conditions
4JC . . . Critical pressure conditions
4K . . . Constant pressure, isobaric conditions
4L . . . Decreasing pressure conditions
4M . . . Increasing pressure conditions
4N . . . Other pressure conditions
. . . . * Add to AY9 4N letters P/S following BSB J if applicable.
4O . . Velocity conditions
4P . . Thermal conditions, temperature conditions
4PC . . . Critical temperature
4Q . . . Constant temperature, isothermal conditions
4R . . . Adiabatic conditions

[Science in general AY2]
[General processes/properties]
[Conditions & parameters & environments of physical events AY94]
[Thermal conditions AY94P]
. [Adiabatic conditions AY94R]

AY9 4S . Decreasing temperature conditions
4V . Increasing temperature conditions
4Y Electric & magnetic conditions
5 . Electric field conditions
6 . Magnetic field conditions
7 . Radiation conditions
8 . Other conditions
B Dimension (general properties)
. * As a qualifying property.
. * Details for AY9 B/AY9 D are taken from the substantive classes for dimensions at B9B/B9D.
. * Add to AY9 letters B/D following B9 if applicable; eg
B2X . Dimensional analysis
BG . Degree, number
. . * For example: many, few, double, very, optimal.
BH . Nodes, points dimensions
BI . Indexes
BK . Similarity parameters
C . Time general properties
CI . . Duration, life
CM . . . Mean life
CQ . . Frequency
CU . . Periodic
D . Space general properties
DE . . Direction
DF . . One-dimensional space general properties
DG . . Distance
DH . . . Mean free path
DJ . . . Width
DK . . . Radius
DP . . Two-dimensional space general properties
DQ . . Three-dimensinal space general properties
DV . . Multidimensional space, hyperspace general properties
G Systems characteristics (properties)
. * As qualifying properties; for the substantive subject of systems, see AYG.
. * Add to AY9 letters G/S following AY; eg
GE . Disturbances, perturbations
GV . Conservation
JV . Shape, configuration
KV . Continuous
KX . Discontinuous, discrete
LR . Linear
M . Non-linear

[Science in general AY2]
[General processes/properties]
. [Conditions & parameters & environments of physical events AY94]
. . [Systems characteristics properties AY9G]
. . . [Non-linear AY9M]

Subjects of scientific enquiry
. * In principle, the end-product of any discipline constitutes the main facet of that subject. In science, the main product consists of the great bodies of information and methods of enquiry into natural phenomena which constitute the special sciences. These will be found in B/G (the natural sciences) and to a lesser degree (since other forms of knowledge also contribute to them) in H/V (social sciences and technologies). Locate here only general studies of the products of science.
AYE . The Sciences
. . * The totality of the different fields of enquiry in science. This class should not be qualified by classes AY2/AYA except for works treating the different sciences as an aggregate (and usually including a comparative element). When in doubt, prefer AY2/AYA.
. . * For interdisciplinary relations between sciences, see AY2 9XG.
. . *Operations*
AYE 3E . . . Classification of sciences, systematics of the sciences, taxonomy of the sciences
G . . Empirical sciences (general)
. . . * For empirical science, see AY3 2G.
H . . Fundamental sciences, pure sciences, basic sciences
L . . Natural sciences
. . . * Usually taken to exclude the social sciences.
N . . Nature, The natural world, natural history
. . . * In nearly all the literature using this term, Biology is implied, together with environmental factors important to living things. For this use of the term, see Biology ET.
. . . * Locate here works which are truly general, with mineralogy, earth sciences, astronomy, etc. considered as well as biology.
AYF . . General phenomena (extra disciplinary)
. . . * The fields of scientific enquiry are usually defined by the established sciences (physics, chemistry, etc.), each encompassing a level of organization in natural phenomena (eg energy and matter per se, molecular organization (substances) and so on). But some very general phenomena have no particular place in these, appearing in several or all of the individual sciences; eg properties like form, processes like periodicity, abstract entities like systems. One such phenomenon, that of systems, has achieved the

Science in general AY2]
[Subjects of scientific enquiry]
[The Sciences AYE]
[General phenomena extra disciplinary AYF]

status of an independent field of study, applicable throughout the sciences, natural and social; this is given in detail below.
* If considered completely generally (as, for example, in a work on form in nature and in art) these concepts are treated as general phenomena and located in Class 2. If regarded scientifically, they belong here.
* Add to AYF letters A/Y following 3 in Class 3 Phenomena (notation is provisional until publication of Classes 2/9); eg Complementarities AYF C.

AYF X Systems & communication & control, complex phenomena (general)
AYG . Systems, systemology
. . * A set of interacting units with relationships between them (Miller). An organized or complex whole.
. . * For a prominent applications of the systems approach, see systems engineering U.
AYG 2M . . Mathematics & statistics
2M3 L . . . Models & simulation
2M8 H . . . Optimization (systemology)
2P2 H . . . . Hamiltonian systems (systemology)
2RC N8U . . . Fractals (systemology)
3D . . General systems theory, GST
. . *Operations on systems*
9B . . . Analysis & design of systems, planning of systems
. . . . * Theoretical studies only. For practical studies, see system concerned, especially in Technology U/V.
. . . . * For systems analysis in the narrow sense of a preliminary to operational research or computerization, see latter.
9D . . . Prediction (systems approach)
. . *Properties & processes of systems*
C . . . Systems behaviour (general)
E . . . . Error (systems behaviour), disturbance, perturbation, degradation (systems behaviour)
. . . . . * For error analysis, see AY7 DA (Evaluation & testing); for error detection, see UC (Systems engineering).
F . . . . . Compensation (systems error)
G . . . . . Transients (disturbances), temporary disturbances
H . . . . Environment of system
. . . . . * Set of elements and their properties which are not part of the system but which may affect it.

[The Sciences AYE]
[General phenomena extra disciplinary AYF]
[Systems & communication & control AYFX]
[Systems AYG]
[Properties & processes of systems]
[Systems behaviour general AYGC]
[Environment of system AYGH]

AYG J Wholeness, holistic process
. * For synergism, see AYI R.
K Teleology (systems)
L . Directiveness
M . Goals of system, objectives of system
N . Ideals of system
P State of system
. * Set of properties relevant to the behaviour of a system at any given time.
Q . Reliability
R . Consistency
S . Sensitivity
T . Constancy,
V . . Conservation (systems behaviour)
. . . * For conservation laws, see Physics B9G V.
X . Homogeneity
Y . Heterogeneity
AYH . Variability, variance (general properties)
AYH H . . Events in system, change of state in system
. . . * For conflict and competition, see Biological systems E.
HM . . . Degree of freedom
I . . . Transformation, phase transitions (systemology)
J . . . Periodic change
K . . . Cycles
L . . . Action & reaction (systems)
. . . . * Event caused by another event.
LM . . . . Susceptibility
MP . . . . Persistence
MR . . . . Resistance
N . . . Response (systems)
. . . . * Event is associated with another event.
NT . . . . Time response
NW . . . . . Delay

General processes/properties

[Systems behaviour general AYGC]
[State of system AYGP]
[Variability AYH]
. [Events in system AYHH]
. . [Response systems AYHN]
. . . [Time response AYHNT]
. . . . [Delay AYHNW]

AYH Q . . . . . Hysteresis (systems behaviour)
. . . . . . * Dependence of the state of a system on its
. . . . . . previous history; usually seen in the lagging
. . . . . . of a physical effect behind its cause.
R . . Acts (systems), autonomous events (systems)
. . . * Autonomous changes, self-determined.
. . . * For self-differentiation, see Biological systems, E
S . Variations special to a subject
T Deterministic behaviour, determinate behaviour
TU Probabilistic behaviour
TV Static behaviour
TY Dynamic behaviour
U Stability (systems behaviour)
V . Homeostasis, self-regulation
VV . . Compensation (systems behaviour)
VW . . . Feedback
. . . . * See also feedback control in Cybernetics AYQ J
VX . Asymptotic static stability
W Instability
X . Chaos, catastrophe
. . * A state of disorder whose development in time is
. . highly sensitive to initial conditions, which may lead
. . to sudden and unpredictable departures from
. . equilibrium. Implies some degree of non-linearity.
. . * See also Chaos theory (mathematics)
. . AWE ME3 CC
YC . . Attractors
YE . . Non-attracting sets
AYI C Reversibility
E Irreversibility
H Adaptive behaviour
. * See also Goals AYG M; Adaptive systems AYN P
HK . Other-other (systems behaviour)
. . * Adapts to external change by modifying
. . environment.
HM . Other-self (systems behaviour)
. . * Adapts to external change by modifying self.
HP . Self-other (systems behaviour)
. . * Adapts to internal change by modifying
. . environment.
HS . Self-self

[Systems & communication & control AYFX]
[Systems AYG]
[Properties & processes of systems]
. [Systems behaviour general AYGC]
. . [State of system AYGP]
. . . [Adaptive behaviour AYIH]
. . . . [Self-self AYIHS]

. . . . . * Adapts to internal change by modifying self.
AYI K . . . Growth in time (systems), systems
. . . development
L . . . . Relative growth
. . . . . * Of systems components.
N . . . Finality
. . . . * Direct cause/effect relation between initial
. . . . conditions and final state. Typical of physical
. . . . systems.
P . . . Equifinality
. . . . * Final state may result from different initial
. . . . conditions and by different processes. Typical of
. . . . biological systems. In open systems, this maintains
. . . . the steady state.
Q . . . Merging with other systems
R . . . . Synergy, synergism (general systems)
. . . . . * Additional benefits accrue on coalescence ('the
. . . . . whole greater than its parts').
. . . . . * See also Wholeness AYG J
T . . . . Coupling
. . . . . * See also Man-machine interface, in Class 8
. . . . . Computer science; Symbiosis, in Class E Biology
Subsystems
AYJ . Structure of systems, organization of systems
AYJ H . . Arrangement
J . . . Order
K . . . . Hierarchy (systems)
P . . . . Pattern
S . . . . . Symmetry
T . . . . . . Parity
V . . . . Shape, configuration
W . Networks (systems)
Types of systems
. *By abstractness*
AYK B . . Abstract systems
D . . . Ideal systems, perfect systems
. . . . * Conforming to certain simplifying assumption
. . . . (usually to assist analysis)
F . . . Real systems, non-ideal systems, imperfect
. . . systems
. . . . * Ideal assumptions are qualified by particular
. . . . conditions.
J . . Concrete systems
K . . . Biological models (general systems theory)
. . . . * Use only when taken as models for non-biological
. . . . systems. For biological systems per se, see E; for
. . . . Equifinality, see AYI P.
. *By statistical concepts*
L . . Analytic systems
N . . Stochastic systems

[The Sciences AYE]
[General phenomena extra disciplinary AYF]
[Systems & communication & control AYFX]
[Systems AYG]
[Types of systems]
[By statistical concepts]
. [Stochastic systems AYKN]

AYK O . . Markov processes (systemology)
P . Stationary systems
Q . Non-stationary systems
S . Multivariable systems
V . Continuous systems
X . Discontinuous systems, discrete systems

By composition, structure

AYL D . Ordered systems
F . Disordered systems

By openness

N . Open systems, self-maintaining systems
P . Closed systems

By linearity

R . Linear systems
AYM . Non-linear systems
. . * See also Multivariable systems AYK S; Chaos AYH X
AYM S . . Large-scale systems, complex systems

AYN G *By behaviour*

. * Add to AYN letters GJ/IT following AY for concepts not enumerated at AYN J/AYN P below; eg Holistic systems AYN GJ.

By environment

. * See also open & closed systems AYL N/P
G . Metasystems
. . * Systems over and beyond those of a lower logical order (eg the system of monetary values affecting decisions in a gambling game).

. *By state of system*

J . . Reactive systems
K . . Responsive systems
L . . Autonomous systems
MN . . Deterministic systems
MP . . Probabilistic systems
MQ . . Static systems, one-state systems
MS . . Dynamic systems, multi-state systems, non-equilibrium systems
. . . * State changes over time.
MT . . . Simple dynamic systems
. . . . * For example, clockwork systems.
NS . . Stable systems
O . . . Self-regulating systems, self-organizing systems, homeostatic systems

[Systems & communication & control AYFX]
[Systems AYG]
[Types of systems]
[By environment]
. [By state of system]
. . [Stable systems AYNNS]
. . . [Self-regulating systems AYNO]

. . . . * Elements & environment are dynamic.
AYN P . . Adaptive systems
. . . * From Ackoff's categorization.
PR . . . State-maintaining systems
PS . . . . Variety-increasing systems
PT . . . Goal-setting systems, goal-seeking systems
PU . . . . Multi-goal-seeking systems
PUW . . . . . Purposive systems
PV . . . Purposeful systems
PVW . . . . Ideal seeking systems

By function

. * Characterized by a special function rather than by structural or behavioural features.
. * See also Natural systems (B/K), Technological systems (U/V) and other classes to which systems theory may be applied.
. * An alternative (not recommended) for libraries wishing to keep together works dealing with particular fields from a systems point of view is provided here. If this option is taken, proceed as follows:
AYO . Applications of systems theory
. . * Alternative (not recommended) to subordinating to subject.
. . * Add to AYO numbers and letters 3/9,A/Z from the whole classification.

AYP M . Communication & control (systems theory)
. . * For control systems in technology, see U7.
N . . Information theory, communication theory (information theory)
P . . . Transmission & reception (information theory)
P3J . . . . Shannon's theory
Q . . . . Transmission (information theory)
R . . . . Reception (information theory)
S . . . Signalling
T . . . . Coding
TU . . . . . Decoding
TV . . . . . . Decodability
TW . . . . . Correction codes
TX . . . . . Group codes
U . . . Channel capacity (information theory)
V . . . Redundancy
W . . . Interference (information theory)
X . . . . Noise (information theory)

[Systems AYG]
[Types of systems]
[By function]
[Communication & control systems theory AYPM]
[Information theory AYPN]
. [Interference information theory AYPW]
. . [Noise information theory AYPX]

AYQ Cybernetics, control (systems theory)
. * Definitions vary; in the West, it generally refers to the scientific theory of communication and control. Elsewhere, it often refers to control and information processing together (including computers).
. * See also Control systems in technology U7

AYQ 2M . Mathematics & statistics
2M8 YT . . Transfer function
2WK . . Optimization (cybernetics)
. Theory
8F . . Control theory
. *Operations on system*
. . * Add to AYQ 9 letters B/D following ATG 9; eg
9B . . Analysis & design
. *Properties & processes*
. . * Add to AYQ letters C/G following AYG;
. . * Add to AYQ H letters H/J following AY; eg
GR . . Consistency
GS . . Sensitivity
HHU . . Stability
HIH . . Adaptive behaviour
. *Subsystems*
. *Types of cybernetic systems*
. . *By input-ouput relation*
I . . . Open loop systems, environmental control
. . . . * No direct control.
J . . . Closed loop systems, feedback control systems, negative feedback control systems
. . . . * For automata, see Articial intelligence AYR.
. . *By general systems theory concepts*
. . . * Add to AYR letters K/P following AY, with the additions indicated.
AYR KN . . . Stochastic control system
KO . . . . Markov processes control systems
KP . . . Stationary control systems

[Types of systems]
[By function]
[Communication & control systems theory AYPM]
[Cybernetics AYQ]
. [Types of cybernetic systems]
. . [By general systems theory concepts]
. . . [Stationary control systems AYRKP]

AYR KQ . . . Non-stationary control systems
KS . . . Multivariable control systems
KV . . . Continuous control systems
KX . . . Discontinuous control systems, discrete control systems
LL . . . Integral control systems
LR . . . Linear control systems
LS . . . . Bilinear control systems
M . . . Non-linear control systems
MS . . . Dynamic systems
MT . . . Optimal control systems
NP . . . Adaptive control systems
O . . . Self-organizing control systems
P . . . Fuzzy control systems
S . . *Other types of control systems, A/Z*
. . . * Arrange A/Z: eg proportional, derivative, time-sharing.

AYS Artificial intelligence, machine intelligence
. * Branch of computer science which assumes that computers can be programmed to behave as though they are exercising human intelligence. Embraces pattern recognition, knowledge-based systems, automata and robots.
. * Alternative (not recommended) to subordinating to computer science in Class 8.

[The Sciences AYE]
[General phenomena extra disciplinary AYF]
. . [Systems AYG]
. . . [Types of systems]
. . . . [By function]
. . . . . [Communication & control systems theory AYPM]
. . . . . . [Artificial intelligence AYS]

AYY Physical sciences (general)

Outline of Physics

B PHYSICS
B2 . *Common subdivisions*
. . * As AY2; eg
B29 A . Social aspects of physics
X . Physics as a discipline... Philosophy...
B2M . Mathematical physics... Statistics in physics
. . * As AM.
. *Operations & agents of operations*
B32 . . Research in physics (general)
B32 C . . . Methodology...
. . . *Types of investigation*
B36 . . . . Practical physics
. . . . . * As for Practical science AY36/7, with
. . . . . adjustments.
B37 . . . . . Unwanted effects... Safety
. . . . . precautions...
B3B . . . . . Equipment & materials
B3B D . . . . . . *Operations on... Handling...*
B3R . . . . . . Materials (general)
B3U . . . . . . Equipment & plant (general)
B4 . . . . . . Instrumentation (general)
B45 . . . . Instrument components
B5 . . . . *Types of instruments*
B62 . . . . . Investigative techniques in physics
. . . . . . *Serving all techniques & objectives*
B63 . . . . Data processing & recording...
. . . . . . *By scale*
B67 . . . . Microtechniques...
B69 . . . . . . Physical methods, physical techniques
. . . . * As BA/BW, with amendments & additions.
B6B . . . . Mechanical... Vibration techniques...
B6H . . . . Electromagnetic & electrical
. . . . techniques
B6H I . . . . Electrial techniques
B6I B . . . . Electronics
. . . . * Divided as in Technology U.
C . . . . Semiconductors...
P . . . . Superconductors...
B6J . . . . Magnetic techniques
B6K . . . . Radiation, wave & pulse techniques
B6K Q . . . . Microwave & optics (together),
. . . . quantum optic techniques
R . . . . Masers... Lasers...
TX . . . . Electrical engineering techniques
U . . . . Telecommunications
. . . . techniques...
V . . . . Microwave techniques... Radar...
B6L . . . . Optical techniques...
B6L WY . . . . Radiological techniques
X . . . . X-rays... Gamma rays...
B6M . . . . Particle physics techniques...
B6N P . . . . Electron techniques...
B6O . . . . Nuclear techniques...

[PHYSICS B]
[Operations & agents of operations]
. . [Types of investigation]
. . . [Practical physics B36]
. . . . [Investigative techniques in physics B62]
. . . . . [Physical methods B69]
. . . . . . [Electromagnetic & electrical techniques B6H]
. . . . [Nuclear techniques... B6O]

B6R . . . . . . Bulk matter techniques
B6R GH . . . . Acoustic techniques... Ultrasonic...
GP . . . . Thermal techniques... Cryogenic...
NR . . . . Phase translation techniques
TCW . . . . Gas dynamic... Hydrostatic...
B6X . . . . Chemical techniques...
. . . . . *Techniques by action on phenomena*
B73 . . . . . . Production techniques... Observing...
B76 . . . . . . Measurement... Counting...
B7A . . . . . . Testing & evaluation...
B7C . . . . Monitoring...
B7F . . . . . . Simulating, modelling...
B7H . . . . . . Visualizing & imaging techniques
B7J . . . . Microscopy... Holography...
B7K T . . . . Photography... Radiography...
B7M . . . . . . Spectroscopy, spectrography
B7N . . . . Spectrometry
. . . . * If distinguished from spectroscopy.
B7P . . . . . . Tracer techniques... Isotopes...
. . . *Investigations by special environment*
B7X . . . . Vacuums... Subsurface... Space...
B7X N . . . . Non-experimental
B82 . . . . Experimental research
. . . . . * If distinguished from practical physics.
. . . *Investigations by objective*
B85 . . . . Fundamental research... Oriented...
. . . *Investigations by methods, techniques*
B8B . . . . Theoretical physics (general)
. . . . . Particular theories
. . . . . . * Theories restricted to a specific field go
. . . . . . with that field.
B8D . . . . . . Classical, Newtonian... Statistical
. . . . . . mechanics...
B8F . . . . . . Field theory (general)
B8H . . . . . . Relativity theory (general)
B8M . . . . . . Quantum theory (general)
B8M N . . . . Quantization, quantum numbers
. . . . (general)
B8P . . . . . . Wave mechanics... Matrix
. . . . . . mechanics...
B9 *General processes & properties*
. * B92/99 are for use only as qualifiers.
. * As AY92/99.
B92 . Distribution... Change.. Cyclical change...
B94 . *Conditions, parameters, influences*
B94 K . . Isobaric... Isothermal...
B9B . Physical dimensions... Similarity...
B9C . . Time... Horology... Duration...
B9D . . Space... Size...

Outline of Physics

[PHYSICS B]
[General processes & properties B9]
. [Physical dimensions... Similarity... B9B]
. . [Space... Size... B9D]

B9D F . . . One-dimensional... Radius... 3-d,
. . . bodies...
B9G . Systems characteristics
. . * As AYG/R Systems theory.
B9G V . . Conservation laws (general)
B9J . . Structure... Symmetry...
B9K V . . Continuous systems... Non-linear
. . systems...
BAE Energy & matter (together)
BAE 9GW . Equivalence of energy & matter
BAF Energy interactions & forms
BAG . Thermodynamics
. . * Interactions between energy systems and their
. . effect on the states of the systems. For thermal
. . phenomena narrowly, see Bulk matter physics
. . BRG P.
BAH . . Principles & laws of thermodynamics
. . *Properties & processes*
BAP J . . . Internal energy... Free energy...
BAT . . . Transport processes (general)...
BB . Mechanics
BB8 B . . *Theory*
. . . * Classical, relativistic, quantum mechanics, etc
. . . are treated as theories of general physics and are
. . . preferred at B8B. This location is an alternative.
BBB . . Energy... Potential energy... Kinetic...
BBG . . Force(s)... Moment, torque...
. . . *Types of forces*
BBJ . . . . Pressure...
BBK . . . . Deformation... Stress... Strain...
. . . . . * The detailed schedule is at BRB K under
. . . . . bulk matter (which is usually implied).
BCB . . . . Elasticity
BCE L . . . . Couple... Attraction.. Repulsion...
BCF . . . *Forces special to a context*
. . . . * Eg, energy loss (particles) BMC F.
BCH . . Statics
BCI . . . Inertia... Mass... Density...
BCN . . . Equilibrium... Stability... Instability...
BCS . . Motion... Momentum...
BCX . . Dynamics
BDA . . . Kinematics, pure motion...
BDB . . . . Displacement...
BDC . . . . Velocity... Acceleration...
BDE . . . Kinetics
. . . *Forms of motion*
BDK B . . . . Of points... Of extended figures...
BDM . . . . Circular motion... Rotation... Spin...
BDP . . . . . Vortices...
BDS . . . . Periodic motion... Harmonics...
BE . . . . . Oscillation & vibration
. . . . . . * Add to BE letters A/V following BF; eg
BEB . . . . . . Harmonics...

[PHYSICS B]
[Energy interactions & forms BAF]
. . [Dynamics BCX]
. . . [Forms of motion]
. . . . [Periodic motion... Harmonics... BDS]
. . . . . [Oscillation & vibration BE]
. . . . . . [Harmonics... BEB]

. . . . . . *Types of oscillation & vibration*
BEX C . . . . *By degree of freedom*
. . . . *By external/internal origin*
H . . . . Forced vibration...
BEY . . . . . Radiation (general)
. . . . . . * If distinguished from wave motion (BF). For
. . . . . . electro-magnetic radiation, see BK.
BF . . . . . Waves, wave motion
. . . . . . * Add to BFA letters A/E following B.
. . . . . . *Properties & processes*
BFB . . . . Harmonics (waves)
BFC . . . . Propagation, transmission...
BFD . . . . Frequency... Spectra...
BFF . . . . Coherence... Emission...
BFL . . . . Absorption... Refraction...
BFN . . . . Reflection... Resonance...
BFP . . . . Polarization...
BFQ . . . . Diffraction... Interference...
BFS . . . . Collision...
BFT . . . . Scattering... Diffusion (waves)...
. . . . . . *Types of radiation/wave motion*
BFV V . . . . *By property*
. . . . * As BF; eg polarized waves BFV VP.
. . . . *By directional & transience factors*
BFW C . . . . Isotropic waves... Waveforms...
BFY G . . . . Standing waves... Continuous...
R . . . . Transverse waves (general)...
U . . . . Longitudinal waves (general)...
BGA . . . . Shock waves... Blast waves...
BGB . . . . Beams... Rays...
BGH . . . . Acoustic properties... Thermal properties...
. . . . . * See Bulk matter, BRG H.

Special energy forms
BGR . Gravitation... Gravitational waves...
BGR G . . Ballistics...
BGY . Electricity & magnetism...
. . * The main schedule for this class is under bulk matter
. . at BRG Y. All the details there are available here.
BH . . Elecromagnetism... Electromagnetic field...
BHI . . Electricity, electrical properties
BHK . . . Charge... Voltage...
BHN . . . Electrostatics... Electrodynamics...
BHP . . . Current... Circuits... Conduction...
. . . . * For semi-conductors, see solid state BVI; for
. . . . electronics in general, see Techniques B6I B.
. . . *Interactions with other energy forms*
BIU B . . . . Mechanoelectric effects... Photoelectric
. . . . effects...
BJ . . Magnetism

Outline of Physics

[PHYSICS B]
[Special energy forms]
. [Electricity & magnetism... BGY]
. . [Magnetism BJ]

BJK . . . Magnetic flux... Magnetization...
BJQ . . . Diamagnetism... Paramagnetism...
BK . . Electromagnetic radiation
BKF C . . . Propagation, transmission
P . . . Polarization
. . . *By product*
BKJ . . . . Ionizing radiation... Non-ionizing...
. . . *By wavelength & frequency*
BKM . . . . Radiofrequency waves, Hertzian waves...
BKQ . . . . Microwave & optics (together)
. . . . . * For masers and lasers, see stimulated emission
. . . . . of radiation (techniques) B6K R.
BKU . . . . Microwaves... UHF... SHF...
BL . . . . Optics
. . . . . * The main schedule is under bulk matter, at
. . . . . BRL. All the details there are available here.
. . . . . *Properties & processes*
BLF . . . . . . Wave motion Physical optics...
BLF C . . . . . . Propagation, transmission
BLG C . . . . Rays, geometrical optics...
BLH . . . . . . Electrooptics...
. . . . . *Special properties & processes*
BLL . . . . . . Luminosity... Colour...
. . . . * See Bulk matter BRL L
BLP . . . . . *Types of light by wave property*
. . . . . . * As BF; eg
BLP F . . . . . . Coherent light...
. . . . . *By wavelength & frequency*
BLU . . . . . . Infrared radiation...
BLV . . . . . . Visible light... Ultraviolet...
BLW Y . . . . Radiology
BLX . . . . . X-rays... Gamma rays...

BLY Matter

BM Particle physics, nuclear physics (broadest sense)
. * For elementary particles specifically, see BNB.
BM3 6 . Practical & experimental, high energy physics
. . *Investigative operations*
BM7 3 . . . Particle production... Separation...
4G . . . Detecting & indicating...
5 . . . Counting & detecting...
. . . . Instruments
54 . . . . . Ionization counters... Scintillation
. . . . . counters...
H . . . Track visualization... Cloud chambers...
IM . . . Beam handling, particle optics...
T . . . Acceleration... Accelerators... Cyclotrons...
BM8 B . Theory
FG . . Unification theories... Gauge theories...

[PHYSICS B]
[Particle physics BM]
[Theory BM8B]
. [Unification theories... Gauge theories... BM8FG]

BM8 M . Quantum theory
MJF . . Relativistic quantum field theory
Processes & properties
BMA F . Energy interactions & forms, particle
. interactions
BMB B . . Energy
C . . . Energy ranges... Low... High energy...
D . . . Energy levels & states
E . . . . Stationary state... Excited state...
EM . . . . Bound state...
F . . . Energy bands
. . . . * See condensed matter BTX BF
BMC F . . . Energy loss of particles...
X . . Dynamics
BMD N . . . Rotation...
. . . . * For angular momentum, spin, etc, see special
. . . . quantum properties BMM D.
BMF . . . Wave properties
BMF G . . . . Emission... Decay...
RU . . . . Coupling...
S . . . . Collision... Scattering..
UR . . . . Capture... Annihilation...
BMH . . Electromagnetic properties...
. . . * For charge, see special quantum properties
. . . BMM M.
BMJ . . . Magnetism
. . *Special particle processes & properties*
BMM . . . Quantum number properties
BMM B . . . . Symmetry...
E . . . . . Parity... Spin... Magnetic moment...
RU . . . . . Charm... Strangeness...
. . *Interactions by energy expenditure*
BMN J . . . Superhigh energy...
V . . Basic interactions
X . . . Exchange forces, field particle exchange
BMO . . . . Exchange particles, gauge bosons, field
. . . . particles (general), quasi-particles
BMP G . . . Gravitational interaction
J . . . Electroweak forces
. . . . Elecromagnetic interactions
. . . . . * See charged particles, BNG.
L . . . . Weak interactions
N . . . . Strong interactions
. . . . . * See Hadrons BNQ PN
BMQ . . *Interactions with another particle*
. . . * This class appears only under specific particles as
. . . required and is divided like BN. See BNQ for
. . . examples.
BMV *Parts of particles*
BMW . Composite particles, composite models...
BNB *Types of particles Elementary particles types*

[PHYSICS B]
[Particle physics BM]
[Types of particles Elementary particles types BNB]

* For exchange particles, see BMO.
* Each type of particle is divided as follows (where - represents the particle's classmark):
* Add to - letters A/Q following BM (for Processes & properties);
* Add to -QY letters V/Y following BM (for Parts);
* Add to -R letters B/X following BN (for types by other particles); eg BPR Q Hadronic atoms;
* Add to -S letters O/Q following B (continuing types by other particles);
* Use letters T/Y for types special to a given particle; eg BNT T Dibaryons.

BNC *By quantum property*
. * As BMM; eg
BNC AS . Conserved particles...
By source
BND A . Particle asscelerator particles...
C . Cosmic rays...
By aggregation
RD . Many-particle systems...
By lifetime
RH . Stable particles... Resonances...
By relation between quantum numbers
BNF . Antiparticles...
By charge
BNF X . Electromagnetic field particles
BNG . . Electromagnetic interaction (particles)
. . . Exchange particles
BNG O . . . . Photons...
BNH . . Electrically charged particles...
BNI . . Magnetically charged, magnetic monopoles
By spin
BNJ . Fermions... Bosons...
By energy characteristics
BNL C . Monoenergetic... High energy particles...
By mass
Q . Light particles... Heavy particles...
By strength of interaction
BNM . Leptons
BNN . . Muons... Neutrinos...
BNP . . Electrons... Positrons...
BNQ . Hadrons
BNQ MB . . Symmetry
MCS . . . Unitary symmetry...
MRW . . Strangeness...
PN . . Strong interactions (general)
Q . . *Interactions with other particles*
QP . . . Electron-hadron interactions
. . *Types of hadrons*
T . . . Multiplets... SU3 groups...
BNR . . . . Quarks
. . . . . *Properties*
BNR MRS . . . . . . Colour... Charm...

[Particle physics BM]
[Types of particles Elementary particles types BNB]
. . . [Types of hadrons]
. . . . [Multiplets... SU3 groups... BNQT]
. . . . . [Quarks BNR]
. . . . . . [Properties]
. . . . [Colour... Charm... BNRMRS]

. . . . . . *Types of quarks*
BNR RF . . . . Antiquarks... Flavours...
BNS . . . . Mesons
BNS T . . . . . Pions... Kaons... Eta-mesons...
BNT . . . . Baryons
BNU . . . . . Nucleons... Protons... Neutrons...
BNX . . . . . Hyperons... Lambda particles...
BNX T . . . . . . Lambda particles... Sigma particles
BNY Atomic & nuclear physics (together)
BO . Nuclei, nucleus, nuclear physics
. . *Operations*
BO7 H . . . Track visualization
. . Theory, models
BO8 VC . . . Unified model... Liquid drop model...
. . *General properties*
BO9 JV . . . Shape, configuration
BOA F . . Energy interactions & forms
. . . * For nuclear reaction in the narrower sense (usually assumed) see BOR.
BOB B . . . Nuclear energy
D . . . . Nuclear energy levels...
DN . . . . . Transitions
G . . . Nuclear forces
GH . . . . Binding energy...
BOC X . . . Dynamics
. . . . Radiation
BOF J . . . . . Decays
K . . . . . . Radioactivity... Beta decay...
S . . . . . Collisions... Scattering...
BOH . . . Electromagnetism
. . . . * For nuclear magnetic moment, see BOM NJ
. . *Special nuclear processes & properties*
. . . *Quantum number properties*
BOM B . . . . Symmetry... Parity...
NJ . . . . Nuclear magnetic moment
O . . . . . Nucler magnetic resonance, NMR...
BON V . . . Basic interactions
BOQ . . . *Interactions with other particles*
BOQ M . . . . Lepton-nucleus reactions
U . . . . Nucleon-nucleus reactions
BOR . . . Nuclear reactions
. . . . * Between nucleus and bombarding particles.
. . . . * The provisions indicated at BNB for the subdivision of any particle are modified here in order to provide more room for special nuclear reactions. Types of nuclei follow at BOX D.
. . . . * For constituents of reactions (collision, scattering, etc.) see BOF.
BOR BB . . . . Nuclear reaction energy, Q-factor
. . . . *Types of nuclear reactions*

Outline of Physics

[PHYSICS B]
[Particle physics BM]
[Atomic & nuclear physics together BNY]
[Nuclei BO]
. [Special nuclear processes & properties]
. . [Nuclear reactions BOR]
. . . [Types of nuclear reactions]

BOR PN	. . . . Natural... Artificial...
Q	. . . . Chain reaction... Cyclic...
BOS	. . . . *By product nucleus*
BOS HD	. . . . . Deuteron... Helium-4...
	. . . . *By emitted radiation/particle*
T	. . . . . Gamma particle product reaction...
BOT	. . . . *By projectile, incident radiation*
BOT HD	. . . . . Deuteron... Alpha particle...
BOU	. . . . *By incident & emitted radiation/particle combined*
BOU V	. . . . . Proton (incident & emitted)
Y	. . . . *By target nucleus*
	. . . . . * Add to BOV letters D/R following BOX; eg
BOV HD	. . . . . Deuterons (targets)... Lithium...
	. . . . *By mode of energy release*
BOW	. . . . . Fission
BOW 3W	. . . . . . Reactors
BOX B	. . . . . Fusion
B3W	. . . . . . Fusion reactors
B7R	. . . . . . Plasma confinement...
	. *Types of nuclei*
FQ	. . Active... Unstable, radionuclides
	. . *By mass number*
GB	. . . Nuclei with mass numbers 1-5...
	. . *By host chemical element*
HB	. . . Hydrogen... Deuterons... Tritons
HH	. . . Helium nucleus, alpha-particles...
BOY	Atoms, molecules & ions (together)
BOY BD	. Energy levels
NV	. Basic interactions
BP	. Atomic physics, physics of single atoms
BPB B	. . Energy
D	. . . Energy levels... Transitions...
DT	. . . . Atomic orbitals...
EU	. . . . Multiplets... Triplet state...
BPE Y	. . Radiation
BPF S	. . . Collisions...Scattering...
BPG B	. . . Atomic beams
	. . *Interactions with other particles*
BPQ P	. . . Electron-atom interaction
YP	. . . Atom-atom interaction
	. . *Parts*
	. . . * See Electrons BNP; Nucleons BNU; Nuclei BO
	. . *Types*
	. . . * For ions, see BQU.
	. . . *By mass*
BPR LS	. . . . Heavy atoms...
BPV	. . . Nuclides... Isotopes...

[PHYSICS B]
[Particle physics BM]
. [Atomic & nuclear physics together BNY]
. . [Atoms BOY]
. . . [Atomic physics BP]
. . . . [Types]
. . . . . [Nuclides... Isotopes... BPV]

BQ	. . . Molecules (physics), physics of single molecules
	. . . . Energy levels
BQB D	. . . . . Molecular electronic structure
DQ	. . . . . . Isomerism... Molecular orbitals...
BQC F	. . . . . Energy loss... Molecular stopping power...
BQN V	. . . . Basic interactions
BQP L	. . . . . Weak interactions
LBG	. . . . . . Van der Waals forces
	. . . . *Interactions with another particle*
BQQ W	. . . . . Neutron-molecule interaction
	. . . . *Types of molecules*
BQT D	. . . . . Diatomic molecules... Conjugated...
BQU	. . . Ion physics
	. . . . Radiation
	. . . . . * For ionization, see BKJ.
BQU F	. . . . . Ion optics
	. . . . . *Interactions with other particles*
QYP	. . . . . . Atom-ion interaction...
QYQ	. . . . . . Molecule-ion interaction...
	. . . . *Types of ions*
	. . . . . *By charge state*
RHU	. . . . . . Cations... Anions... Paired ions...
	. . . . . *By origin*
TIU	. . . . . . Metastable ions... Recoil ions...
BQX	Vacuums
BR	Bulk matter physics, macrophysics
	. Energy interactions & forms
BRA G	. . Thermodynamics
	. . . * For thermal properties narrowly, see BRG P.
T	. . . Transport properties
BRB	. . Mechanics
BRB G	. . . Forces
K	. . . . Deformation
	. . . . . * The full schedule for this is at BVB K.
BRC X	. . . Dynamics
BRE	. . . . Mechanical vibrations, sonics
BRG H	. . . . . Acoustics, sound... Ultrasonics...
HN	. . . . . . Noise
N	. . . . . . Ultrasonics...
P	. . . . . Thermal properties...
Q	. . . . . . Heat transfer... Heat loss...
V	. . . . . . Temperature...
	. . . . . . Thermal regimes
W	. . . . Low temperature physics, cryogenics

Outline of Physics

[Bulk matter physics BR]
[Energy interactions & forms]
. [Mechanics BRB]
. . . [Mechanical vibrations BRE]
. . . . [Thermal properties... BRGP]
. . . . . [Thermal regimes]
. . . . . . [Low temperature physics BRGW]

BRG X . . . . . . High temperature physics
Y . Electrical & magnetic properties
. . * The full schedule for this class is at BVG Y; all its
. . details are available here.
BRL . Optics (bulk matter)
. . * This is the main schedule for optics; all its details are
. . available at BL and in BS/BW.
BRL 4 . . Optical instruments
F . . Physical optics
FH . . Luminescence... Transparence...
GH . . Acoustooptics... Electrooptics...
L . . Luminosity... Colour...
. . *Types of light*
PP . . . Polarized light... Optical solitons...
U . . . Infrared light
V . . . Visible light
W . . . Ultraviolet light
X . X-rays
BRM . Particle & high energy physics in bulk matter
. . * Add to BRM letters M/Q following B; eg
BRM O . . Nuclear physics of bulk matter... Atomic...

BRN States of matter, physical phases (states)
. * For chemical studies, see Chemistry C .
. *Processes*
BRN P . . Change of state
R . . . Phase transformation, phase transition
S . . . . Latent heat... Phase equilibrium...
BRO . . . *By states involved*
. . . . * Use only when qualifying a particular state of
. . . . matter; eg BUO G Liquids - Evaporation
. . . . * When two different states interact, the denser
. . . . state is cited first and classmarks are built
. . . . retroactively; eg liquefaction of gases goes under
. . . . liquids, at BUO H.
. *Subsystems*
BRQ . . Surfaces... Interfaces... Films...
. *Systems, by number of components*
BRR N . . Single... Mixtures...
. *Systems, by number of phases*
. . * Phase here means a state of matters, not a chemical
. . substance or crystal structure.
BRS N . . Homogeneous...
T . . Heterogeneous... Dispersions...
BRT V . . . Colloids... Composite materials...

[PHYSICS B]
[Bulk matter physics BR]
[States of matter BRN]
[Systems]
. [Heterogeneous... Dispersions... BRST]
. . [Colloids... Composite materials... BRTV]

Systems by particular state of matter
BRU . Plasmas & fluids (together)
BRU J . . Magnetohydrodynamics, MHD
BRV . Plasmas
BRV 73 . . Plasma generation
73M . . . Excitation, ionization... Confinement...
B . . Flow of plasmas (general)
. . . * See note re amended notation following BSF.
MGY . . Electric & magnetic properties
MJ . . . Magnetohydrodynamics of plasmas
. . *Subsystems*
PW . . . Impurities in plasmas...
. . *Types of plasmas*
SN . . . Homogeneous... Heterogeneous
VB . . . Plasmoids... Collisionless plasmas...

BS . Fluids
BSA G . . Thermodynamics
. . . * For thermal properties narrowly, see BSM GP.
T . . . Transport properties... Diffusion...
W . . . Viscosity, fluidity
BSB . . Fluid mechanics, flow of fluids
BSB 9BK . . . Similarity parameters, dimensionless
. . . numbers
9BM . . . . Mach number... Reynolds number...
G . . . Forces
. . . . * Many of these are mainly associated with, or
. . . . even special to, gases rather than liquids; see
. . . . BTB G for examples.
IQ . . . . Pitching... Rolling... Yaw...
J . . . . Pressure
JW . . . . External forces(
L . . . . . Stress... Loading...
QU . . . . . Buffeting... Drag...
BSC H . . . Statics
P . . . . Stability
PS . . . . . Static stability... Dynamic...
X . . . Dynamics
BSD E . . . . Kinetics
. . . . *Forms of motion*
N . . . . . Rotational flow... Vortices...
BSF . . . . . Waves... Shock waves...
. . . . *Special attributes & elements of flow*
. . . . . * Normal retroactive synthesis is interrupted
. . . . . here to accommodate the large vocabulary
. . . . . of fluid flows. It is resumed at BSM.
BSG FC . . . . . Suction... Wake...
L . . . . . Jets... Cavities... Boundary layers...
. . . . *Types of flow*

Outline of Physics

[Bulk matter physics BR]
[States of matter BRN]
[Systems by particular state of matter]
[Fluids BS]
[Fluid mechanics BSB]
. [Dynamics BSCX]
. . [Types of flow]

BSG RE . . . *By compressibility Compressible...*
RJ . . . *By viscosity Viscous flow...*
. . . *Inviscid...*
SM . . . *By degree of attachment Separated*
. . . *flow...*
U . . . *By pressure Isobaric flow...*
BSH L . . . *By velocity potential Potential*
. . . *flow...*
. . . *By degree of mixing*
BSI . . . . Laminar flow... Turbulent flow...
BSJ R . . . *By speed*
. . *Flow defined by bodies in fluid*
BSK B . . . Bodies in fluid flow
BBG . . . . Control forces... Longitudinal
. . . . controls...
. . . . *By part of body*
D . . . . . Surfaces... Corners... Holes...
. . . . *By dimension of body*
H . . . . . Two-dimensional...
K . . . . . . Nozzles... Conduits...
Q . . . . . Three-d.... Bodies of
. . . . . revolution...
BSM . *Other energy interactions*
. . * Normal retroactive synthesis is resumed here
. . after its interruption at BSG F.
. . * Add to BSM letters GF/Q following BR; eg
BSM GH . . Acoustic properties
GP . . Thermal properties
M . . *Relations with particles & atomic*
. . *physics*
P . . . Atoms
PBF . . . . Superfluids...
BSN P . . Change of state
Subsystems, parts
BSQ . Surfaces... Interfaces
Types of fluids
BSR D . Perfect fluids...
. . * For superfluids, see BSM PBF.
. *By number of phases*
BSS T . . Heterogeneous... Dispersions...

BT . Gases
BTB . . Gas mechanics, flow of gases
. . . Equipment
BTB 3YE . . . . Wind tunnels...
. . Similarity parameters
9BM . . . Mach number... Reynolds number...
G . . Forces
IQ . . . Pitching... Rolling... Yaw...
JW . . . External forces(

[States of matter BRN]
[Systems by particular state of matter]
[Fluids BS]
[Types of fluids]
[Gases BT]
[Forces BTBG]
. [External forces BTBJW]

BTB Q . . Loads
QU . . . Buffeting... Lift.. Drag...
BTC B . . Aeroelasticity general
X Aerodynamics (narrowly)
BTD V . Vibrations
BTE H . . Flutter... Buzz...
BTG A . . Shock waves...
Special elements & attributes of flow
. * Normal retroactive synthesis is
. interrupted here to accommodate the large
. vocabulary of gas flows. It is resumed at
. BTM.
FC . Suction...
FE . Aeronautical factors
. . * Alternative (not recommended) to
. . locating in V Technology.
FH . . Balancing...
FP . . Manoeuvrability forces...
. . Downwash...
GQ . Slipstream... Wake... Jets..
P . Boundary layer...
Types of gas flow
RE . Compressible flow
BTI . Laminar flow... Turbulent flow...
BTJ S . Subsonic flow... Supersonic...
. *Defined by bodies in flow*
BTK H . . Two-dimensional... Nozzles...
Q . . Three-d.... Bodies of revolution...
X . . *By aeronautical structure*
. . . * See Aeronautical engineering U; eg
. . . fixed wing, sweepback.
BTM *Other energy interactions*
. * Normal retroactive synthesis is resumed
. here after its interruption at BTG F.
. * Add to BTM letters GH/Q following BR;
. eg
BTM GH . Acoustic properties of gases...
GHT . . Sonic boom...
GY . Electrical & magnetic properties
HR . . Electric discharge... Arc
. . discharge...
BTN P . Change of state
Subsystems, parts
BTQ . Surfaces... Films...
Types of systems
. *By number of phases*
BTR D . . Perfect gases... Real gases...
BTT . . Dispersions
. . . * Gases in liquids and in solids go
. . . under liquids and solids.

Outline of Physics

[Systems by particular state of matter]
[Fluids BS]
. [Types of fluids]
. . [Gases BT]
. . . [Types of systems]
. . . . [By number of phases]
. . . . . [Dispersions BTT]

BTU B . . . . . . Plasmas in gases, gaseous plasmas
G . . . . . . Gases in gases
. . . . *By constitution*
. . . . . * For gases of particular elements and
. . . . . compounds, see Class C Chemistry; for air,
. . . . . see Earth's atmosphere DSQ.
. . Liquids
. . . * See under Condensed matter BTX.

BTX Condensed matter physics (narrowly)
. * In present usage, this means liquids and solids.
. together. The term condensed matter is sometimes
. used to mean all forms of bulk matter, in which case
. BR should be used.
BTX BF . Energy bands, band structure...
BFO . . Valence band...
BG . Forces
. . Pressure
BJQ . . . High pressure physics
. . Deformation
BK . . . Rheology
GY . Electrical & magnetic properties
IP . . Superconductivity (general)
. *Parts*
Q . . Surfaces... Interfaces...

BU . Liquids
. . * Add to BU letters A/Q following BS Fluids.
. . Mechanics
BUB . . Hydromechanics, flow of liquids
. . . Statics
BUC H . . . . Hydrostatics
. . . Dynamics
X . . . . Hydrodynamics
BUF . . . . . Waves
. . . *Special attributes & elements of flow*
. . . . * Normal retroactive synthesis is interrupted
. . . . here; it is resumed at BUM.
BUG ER . . . . Flow regimes... Boundary layers...
. . . *Types of flow*
BUI . . . . Laminar flow... Turbulent flow...
. . . . *Flows defined by bodies in liquids*
BUK KN . . . . . Channel flow... Immersed bodies...
BUL W . . . . . Immersed bodies...
. . *Other energy interactions & forms*
. . . * Normal retroactive synthesis is resumed here,
. . . after its interruption at BUG ER.
. . . * Add to BUM letters GH/Q following BR; eg
BUM GH . . . Acoustic properties
M . . . *Relations with particle physics*

[Bulk matter physics BR]
[States of matter BRN]
[Systems by particular state of matter]
[Condensed matter physics narrowly BTX]
[Liquids BU]
. [Other energy interactions & forms]
. . [Relations with particle physics BUMM]

BUN P . . Change of state
BUO E . . . Change of liquid to gas...
G . . . . Evaporation...
H . . . Change of gas to liquid...
I . . . . Condensation...
. *Subsystems*
BUQ . . Surfaces... Films...
. *Types of liquids*
. . *By simplifying assumptions*
BUR D . . . Ideal liquids...
. . *By number of phases*
BUT . . . Liquid dispersions...
BUU G . . . . Gases in liquids...
S . . . . Liquids in liquids...
. . *By composition*
. . . * For liquids defined by chemical
. . . composition, see Class C; for liquid crystals,
. . . see BWU L.

BV Solids, rigid bodies
. * For solid state physics in the narrower sense of
. electric and electronic properties of solids, see
. BVH.
BVB H . Forces
K . . Deformation
KL . . . Stress-strain relationships
L . . . Stress... Tension... Loads...
T . . . Strain
VE . . . . Elastic deformation... Plastic...
YD . . . . Failure... Fatigue...
YK . . . . Creep... Fracture...
BVC B . . Elasticity... Plasticity...
BVG H . Acoustic properties... Thermal...
BVH . Electromagnetism
. . * This is the main schedule for this class, since
. . many of its concepts are dependent on bulk
. . matter and on the solid state in particular.
. . * Many of its concepts (eg circuits) also imply
. . human artefacts and strictly speaking should
. . go in the technology class. Locate works here
. . only if they treat the subject from a strictly
. . physics aspect; in cases of doubt, prefer
. . Technology U/V.
BVH K . . Charge... Voltage...
MC . . Power... Load...
N . . Electrostatics
NBH . . . Electrostatic field
O . . Electrodynamics
P . . . Current
S . . . . Circuits

Outline of Physics

[Systems by particular state of matter]
[Condensed matter physics narrowly BTX]
[Solids BV]
[Electromagnetism BVH]
. [Electrodynamics BVHO]
. . [Current BVHP]
. . . [Circuits BVHS]

BVH SV . . . . Immittance
SX . . . . . Admittance...
U . . . . . . Conductivity, conduction
V . . . . . Impedance... Resistance...
. . . . . . * For dielectrics, see BVI S.
XB . . . *Types of currents*
. . *Electrodynamics of special materials*
BVI . . . Semiconductors
BVI BCX . . . . Electron states... Energy levels...
BF . . . . Energy bands...
E . . . . Charge carriers...
GB . . . . Semiconductor materials
HB . . . . Semiconductor devices
. . . . . * See Semiconductor techniques B6I HB
P . . . Superconductors
S . . . Dielectrics, dielectric materials
SHT . . . . Dielectric strain, displacement...
. *Interactions of electricity with other energy forms*
UB . . Electromechanical effects...
UC . . . Piezoelectricity...
UGH . . Acoustoelectric effects... Thermoelectric effects...
BVJ . Magnetic properties
BVJ FO . . Resonance...
K . . Magnetic flux... Magnetization...
R . . Paramagnetism... Ferromagnetism...
. . *Interactions with other energy forms*
UB . . . Magnetomechanical effects...
UBL . . . . Magnetostriction, piezomagnetism
BVK . . Radiation
. . . * The full schedule for this is at BK; all its details are available here if required.
BVL . Optics
. . * The full schedule for optics is at BRL; all the details there are available here (and at BL if applicable).
BVL L . . Luminosity... Colour...
BVM *Relations of solids with particle physics*
BVN P Change of state of solids
BVO E . To & from gases... Sublimation...
L . To & from liquids... Melting...
Q . . Liquids to solids... Supercooling...
Subsystems
BVQ . Surfaces
BVQ CA . . Tribology, friction... Lubrication...
Systems
. *By number of phases*
BVS N . . Homogeneous
NRS . . . Solid solutions...

[Systems by particular state of matter]
[Condensed matter physics narrowly BTX]
[Solids BV]
[Systems]
[By number of phases]
. [Homogeneous BVSN]
. . [Solid solutions... BVSNRS]

BVS T . Heterogeneous
BVT . . Dispersions... Colloids...
BVU G . . . Gases in solids... Solids in gases...
L . . . Liquids in solids... Solids in liquids...
S . . . Solid in solid systems
Types of solids by composition
. * See Class C Chemistry for metal physics, polymer physics, etc.

BW Crystallography
. * For chemical aspects, see Class C.
. * Alternative (not recommended) is in Class C Chemistry.
BW3 6 . Practical & experimental
BW6 K . . Radiation techniques... Diffraction...
LX . . . X-ray techniques...
BW7 M . . Spectroscopy...
. *Processes & properties*
BWA . . Physical crystallography
BWB K . . . Deformation
XD . . . . Dislocation... Cleavage...
BWL . . Optical properties
BWM . *Relations with particle & atomic physics*
BWN P . Change of state
R . . Phase transitions
. *Processes special to crystallography*
. . * Normal synthesis as instructed at BR is interrupted here. It is resumed at BWR B.
BWO G . . Growth, formation... Nucleation...
J . . Crystallization... Epitaxy...
T . Structural crystallography
BWP . . Lattices
BWP S . . . Symmetry... Defects...
BWQ L . . . Space groups...
. *Subsystems*
V . . Microstructure, grain structure
BWR B . . Surfaces
. . . * Normal synthesis is resumed here after its interruption at BWO.
. *Systems, types of crystals*
M . . Crystal systems
. . . *By symmetry*
BWV N . . . . Body-centred... Face-centred...
. . . *By crystal form*
SD . . . . Pedion... Pinacoid... Dome...
. . . *By bonding*
BWY F . . . . Ionic... Covalent... Metallic...
. . . *By specific element or compound*

[Condensed matter physics narrowly BTX]
[Solids BV]
[Systems]
. [Crystallography BW]
. . [Systems]
. . . [Crystal systems BWRM]
. . . . [By specific element or compound]

BWY P . . . . . Chemical crystallography
. . . . . . * Alternative (not recommended) to locating in
. . . . . . C Chemistry; eg metallic bond crystals
. . . . . . BWY Q.

BX Applied physics, physics-based technologies
. * Alternative (not recommended) to locating in U/V.
. * Add to BX letters following U
. * Add to BY letters following V

Physics

B	Physics
B2	. *Common subdivisions*
	. . * Add to B2 numbers 2/9 in Auxiliary Schedule 1, with the additions and modifications given in AY2 2/9; eg
B22	. . *Forms of presentation*
B23 G	. . . Serials
NL	. . . Tables of constants
P	. . . Technical data
	. . Persons in the subject
B24	. . . Physicists
	. . . . * See also Biography B292
B24 C	. . . . Profession
	. . . . . * For Education, see B26A.
B25	. . Organizations in physics
B25 L	. . Communication & information in physics
B26 A	. . Education, study & teaching of physics
	. . . * Alternative (not recommended) is to locate in JK Education.
C	. . Research
	. . . * This position is used only to qualify classes preceding B32.
B27	. . History of physics
B29 2	. . Biography of physicists
A	. Social aspects of physics, physics & society
X	. Physics as a discipline
	. . * For methodology narrowly, see B32 C (and the notes at AY2 9X).
	. . * Add to B29 X numbers & letters 5/Y following AY2 9X.
	. . * Add to B2 letters A/X following A in classes AA/AX; eg
B2A	. . Philosophy of physics
B2A CKL	. . . Atomism
GHV	. . . The Absolute
GLS	. . . Complementarity
	. . . . * See also Quantum theory B8M P
GM	. . . Cosmology
	. . . . * See also Particle physics theory BM8 B
GQD	. . . Causality
GQE	. . . . Irreversibility (causality)
GQF	. . . . Indeterminacy, uncertainty
GSW	. . . Analogy, similarity
	. . . . * See also Dynamic similarity (dimensions) B9B K
B2M	. . Mathematical methods in physics
B2M 6D	. . . Numerical analysis
7H	. . . Computation
86	. . . Approximation
8L	. . . Functions
8N	. . . . Named functions, A/Z
9L	. . . Equations
9N	. . . . Named equations, A/Z
9P	. . . Inverse problems
B2N B	. . . Non-linear (mathematical physics)
B2P 2	. . . *Systems of individual mathematicians*
	. . . . * For principles, systems, etc. of named persons; eg Hamiltonian principles B2P 2H.
B2R B	. . . Sets
	[Physics B]
	[Physics as a discipline B29X]
	. [Mathematical methods in physics B2M]
	. . [Sets B2RB]
B2R CN8 U	. . . Fractals
B2S A	. . Groups
B2V UYJ R	. . Differential topology
B2W 6Y	. . Variational calculus
G	. . Partial differential equations
IOR	. . Dynamical systems
B2X	. Statistics & probability
B2X 8	. . Statistical methods
FY	. . Probability
N	. . . Stochastic processes
NW	. . . Ergodic processes
O	. . . . Markov processes
	Operations & agents in physics
	. * Add to B numbers & letters 2YM/8 following AY; eg
B2Y Q	. Organization & management of work in physics
B32	. Research operations general
	. . * See note at AY3 2; much of the literature will be at B4/B7, under instrumentation, physical methods, etc.
	. . * Add to B3 numbers & letters 2/X following AY3; eg
B32 B	. . Procedures & methods in physics research
	. . . * For theoretical physics, see B8B.
C	. . . Methodology narrowly
Q	. . . Comparative investigations
B36	. Practical physics
	. . * All procedures other than theoretical physics.
	. . * See also Experimental physics B82 and the general note at AY3.
	. . * Add to B numbers 36/7 following AY.
B37	. . Unwanted effects & safety
B3B	. . Equipment & materials
	. . . *Operations on*
B3C	. . . . Handling techniques
	. . . *Properties*
B3J G	. . . . Accuracy
B3R	. . . Materials in general
B3U	. . . Equipment & plant
B3X	. . . . Laboratories
B4	. . . . Instrumentation
B45	. . . . . Instrument components
B4A	. . . . . . *By energy system*
	. . . . * Details here are taken from Technology U/V (not yet published, so notation is provisional); eg
B4K	. . . . Switching devices, converters
B4L	. . . . Transducers
B5	. . . . . *Types of instruments*
	. . *Operations*

[Physics B]
[Operations & agents in physics]
[Practical physics B36]
[Operations]

B62 Investigative techniques
. *Operations serving all techniques & objectives*
B63 . . Data processing & recording
B64 B . . . Automated methods
D . . . Computer programs
. . . . * Arrange A/Z; e.g. REDUCE B64 DR
. *Operations by scale*
B67 . . Microtechniques

B69 . Physical methods in investigation
. . * Operations acting on physical phenomena, using various physical processes, properties, etc. as agents of investigation.
. . * This class takes works on these techniques only when they are used in physics. Completely general works on them go in AY6 9. When considered as agents in the investigation of a particular problem in physics, class under the problem.
. . * Add to B6 numbers & letters 9,A/W following B in B9/BW with the adjustments indicated (eg at B6I B, which is divided like Technology U, and at B6K T/B6K Y).
B69 B . . Dimensional techniques
B6A T . . Transport techniques (general)
. . . * For heat transfer, see B6R GS
B6B . . Mechanical techniques
B6B IM . . . Torque (techniques)
J . . . Pressure (techniques)
. . . . * For vacuum techniques, see B6Q X.
K . . . Deformation (techniques)
UN . . . . Torsion (techniques)
B6C B . . . . Elasticity (techniques)
E . . . Inertial force (techniques)
EG . . . . Centrifugal force (techniques)
H . . . Static forces (techniques)
K . . . Weight (techniques)
KL . . . . Tare techniques
L . . . Density (techniques)
. . . *Techniques using forms of motion*
B6D C . . . . Velocity (techniques)
CP . . . . . High speed techniques
. . . . . . * In microseconds or shorter.
D . . . . Acceleration (techniques)
LV . . . . Ballistics (techniques)
N . . . . Rotation (techniques), gyroscopic techniques
B6E . . . . Vibration (techniques), oscillation (techniques), sonic (vibratory) techniques
. . . . . * For radiological techniques, see B6L WY.

[Practical physics B36]
[Operations]
[Investigative techniques B62]
[Physical methods in investigation B69]
[Mechanical techniques B6B]
[Techniques using forms of motion]
. [Vibration techniques B6E]

B6G G . . Sonic techniques
H . . Acoustic techniques
. . . * For acoustoelectric techniques, see B6I XR; for thermoacoustics, see B6G PGH; for photoacoustics, see B6L GH; for acoustic holography, see B7K GH
H7S . . . Acoustic signal processing
. . . Emission
HFG . . . . Acoustic emission techniques, acoustic radiation techniques
. . . Absorption
HFL . . . . Acoustic absorption techniques
. . . Resonance
HFO . . . . Acoustic resonance techniques
. . . Interference
HFR . . . . Acoustic interference techniques, acoustic wave interferometry
. . . Scattering
HFT . . . Acoustic scattering techniques
N . . . Ultrasonic techniques
P Thermal techniques
. * Techniques using thermal energy (ie heat possesed by substances, bodies, etc. in the form of molecular vibrational energy).
P3U . Equipment
P3W . . Furnaces (thermal techniques)
PGH . Thermoacoustic techniques
PQ . Quantity of heat
PQ7 6 . . Calorimetry
PQ7 8R . . . Bomb calorimetry
PQ7 8V . . . Microcalorimetry
T . Cooling techniques
U . Heating techniques
. Temperature techniques
V76 . . Thermometry
W . . Low temperature techniques, cryogenic techniques
X . . High temperature techniques, pyrometric techniques
B6H Electromagnetic techniques
B6H I Electrical techniques
. * For electrochemical techniques, see electrolytic techniques, see B6X L.
N . Electrostatic techniques
B6I B Electronic techniques
. * See also particle physics techniques using electrons B7N P.

Investigative techniques

[Operations & agents in physics]
[Practical physics B36]
[Operations]
[Investigative techniques B62]
[Physical methods in investigation B69]
[Mechanical techniques B6B]
[Electronic techniques B6IB]

. * Add to B7I letters following UL Electronic engineering (notation provisional); eg

B6I BM	. Vacuum tube techniques
BS	. Gas discharge tube techniques
C	. Semiconductor techniques
	. . * Add to B6I C letters A/C following BVI.
	. . * Add to BCI D letters D/Q following B if applicable.
	. . * Add to B6I letters E/G following BVI; eg
E	. . Charge carriers
HB	. . Semiconductor devices
HC	. . . Point contact devices
HD	. . . Junction devices, bipolar devices
HE	. . . Diodes (semiconductors)
HG	. . . . Point contact diodes
HJ	. . . . Junction diodes
J	. . . Transistors
JP	. . . . Point contact transistors
K	. . . . Junction transistors
L	. . . . . N-p-n transistors
M	. . . . . P-n-p transistors
N	. . . . . Four layer transistors
O	. . . . Field effect transistors, bipolar transistors
	. . . . . * For those with an optical input and/or output, see Optics BL.
P	. Superconductor techniques
	. . * Add to B6I P letters A/Y following BVI P.
RN	. Electrical resistance (techniques using)
S	. . Dielectric techniques
	. . . * Add to B6I S letters A/M following BVI S.
SLH	. . . Electro-optical techniques
	. *Interactions of electric with other energy forms*
	. . * Add to B6I U letters B/L following BVI U; eg
UB	. . Electromechanical techniques
UC	. . . Piezoelectrical techniques
UGH	. . . Acoustoelectrical techniques
UGP	. . Thermoelectric techniques
UL	. . Photoelectric techniques
B6J	Magnetic techniques

[Operations & agents in physics]
[Practical physics B36]
[Operations]
[Investigative techniques B62]
[Physical methods in investigation B69]
[Mechanical techniques B6B]
[Magnetic techniques B6J]

B6J FO	. Magnetic resonance techniques
	. . * For magnetic resonance imaging, see Imaging B7I O.
FP	. Magnetic polarization techniques
U	. Magnetic field effects (techniques)
UB	. . Magnetomechanical effects (techniques)
UBL	. . . Magnetostriction techniques, piezomagnetism techniques
B6K	Radiation techniques (electromagnetism), wave techniques (electromagnetism)
	. * For tracer techniques, see B7P.
B6K 76	. Measurement
78S	. . Dosimetry
78T	. . . Dosimetry below 5 MeV
78U	. . . . Electron dosimetry
78V	. . . Microdosimetry
FM	. Refraction techniques, refractometry
	. . Instrumentation
FM4	. . . Refractometers
FMF C4	. . Transmission refractometry
FQ	. Diffraction techniques, diffractometry
	. . * Analysis of structure of matter by the diffraction of x-rays, neutrons, etc. in crystals.
	. . * More detail may be taken from crystallography, at BW6 KFQ.
FQ4 FWB	. . Prisms
FQ4 FWG	. . . Diffraction gratings
FQ7 64	. . Diffractometers
	. . *By type of radiation*
	. . . * See the radiation or particle at B6K U/B6Q; eg X-ray diffraction B6L XFQ.
FR	. Electromagnetic wave interferometry
	. . Instrumentation
FR4	. . . Interpherometers
	. . . . * Arrange A/Z; eg
FR5 V	. . . . Etalons
FR5 W	. . . . . Microinterferometers

[Investigative techniques B62]
[Radiation techniques electromagnetism B6K]
[Electromagnetic wave interferometry B6KFR]
. [Instrumentation]
. . [Interpherometers B6KFR4]
. . . [Etalons B6KFR5V]
. . . . [Microinterferometers B6KFR5W]

. *By type of radiation*
. . * See radiation; e.g. acoustic wave interferometry
. . B6R GHF R.
B6K FT Scatter techniques
. * Using irregular reflection, dispersal, etc. of particles
. or waves.
Pulses
FX . Pulse techniques
Beams
GB . Beam handling (general), beam techniques
. (general)
. . * Add to B6K GB letters D/G following AY7 I
. . (Beam handling in imaging); eg
GBF H . . Beam modulation
By result of action
J . Ionizing radiation techniques, radiology
J4U . . Equipment
J4Y . . . Cloud chambers
. . . . * Add to B6K J4Y letters K/P following
. . . . BM7 H; eg
J4Y M . . . Bubble chambers
J4Y P . . . Spark chambers
. . Measurement
J75 4 . . . Counters
J83 . . Production
JP . . Irradiation techniques
JPQ . . . Microirradiation techniques
By wave length/frequency
M . Radiofrequency techniques
Q . Microwave & optical techniques (together)
QM . . Stimulated emission techniques, induced
. . emission techniques, quantum optic
. . techniques
R . . . Maser techniques (general), microwave
. . . amplification by stimulated
. . . emission
. . . . * Add to B6K R letters A/T following B6K S
. . . . (with additions & modifications).
. . . . Instruments
R4 . . . . . Masers
. . . . *Types*
RLP . . . . . Pulsed masers
RST . . . . . Gas masers
S . . . Laser techniques, lasing process, light
. . . amplification by stimulated
. . . emission
. . . . Instruments
S4 . . . . . Lasers
. . . . . . *Components*
S4D S . . . . Cavity resonators

[Investigative techniques B62]
[Stimulated emission techniques B6KQM]
[Laser techniques B6KS]
[Instruments]
. [Lasers B6KS4]
. . [Components]
. . . [Cavity resonators B6KS4DS]

B6K S5 . . *Types of instruments*
. . . * See B6K SL
S7I D Beam handling
S7I FH . Modulating
S7I FW . Pulsing & switching
Processes & properties
SBD . Energy states
SBD L . . Population inversion
. . Excitation
SFI N . . . Stimulated emission of light
SFI O . . . . Laser pumping, optical pumping,
. . . . electronic pumping, pumping
. . . . (lasers)
. Interference
SFR . . Laser interferometry
Types of laser action
. *By energy*
SLH . . High energy lasing process, Q switching
Types of lasers
SLP . Pulsed lasers
SLQ . Continuous lasers
. *By radiation or particle*
. . * Add to B6K S letters LU/Q following B; eg
SLU . . Infrared lasers
SLV . . Visible light lasers
SLW . . Ultraviolet lasers
SO . . Nuclear-powered lasers
. *By medium*
SR . . Lasing action media
ST . . Gas lasers
STV . . . Inert gas lasers
STW . . . Other gas lasers
. . . . * Arrange A/Z.
SU . . Liquid lasers-
SV . . Solid lasers
SVV W . . . Semiconductor lasers, diode lasers
SVX B . . . Ruby lasers
SVX G . . . Glass lasers
SVX J . . . Fibre lasers
SVX M . . . Chemical lasers
. *By action of laser*
SX . . Surface emitting lasers, SEL
. . . * Key component of optic-electric integrated
. . . circuits.
T . . Quantum well lasers

[Investigative techniques B62]
[By wave length/frequency]
[Microwave & optical techniques together B6KQ]
. [Stimulated emission techniques B6KQM]
. . [Laser techniques B6KS]
. . . . [By action of laser]
. . . . . [Quantum well lasers B6KT]

B6K TV . . Non-linear optical techniques
. . . * Effect of very intense light beams on matter
. . . throught which they are propagated. Utilizing the
. . . optical effects of laser radiation interacting with
. . . non-linear materials.
. . . * See also Quantum optic techniques B6K QM
TX Electrical engineering techniques
. * Add details from Class U Technology when
. published.
U Telecommunications techniques
. * Add details from Class U Technology when
. published.
V Microwave techniques
. * For masers, see B6K R.
. * Add to B6K V letters U/W following BK; eg
VVB . Superhigh frequency techniques
VWE . Millimetre wave techniques

B6L Optical techniques
. * For quantum optic techniques see B6K QM; Non-
. linear optical techniques B6K TV; for optical
. techiques in bulk matter, see B6R L.
B6L FC . Optical transmission techniques
FCT . . Fibre optic techniques
FL . Absorption techniques
FM . Refraction techniques
FMJ . . Birefringence techniques, double
. . refraction techniques
. . . * For photoelastic stress analysis, see
. . . BVB KL6 LFM J.
FN . Reflection techniques
FP . Polarization techniques, polarized light
. techniques
FQ . Diffraction techniques (light)
FQS . . Schlieren techniques
FR . Interference techniques, interferometry
FRL . . Interference fringes (techniques)
FT . Scattering (optical techniques)
GH . Acoustooptic techniques, photoacoustics
L . Luminosity (techniques)
M . Colour (techniques)
PF . Coherent light (techniques)
. . * For stimulated emmission of light, see B6K S
. . (lasers)
. *Types of optics by wavelength*
U . . Infrared techniques
V . . Visible light techniques
W . . Ultraviolet techniques

[Investigative techniques B62]
[Mechanical techniques B6B]
[Radiation techniques electromagnetism B6K]
[By wave length/frequency]
. [Optical techniques B6L]
. . [Types of optics by wavelength]
. . . [Ultraviolet techniques B6LW]

B6L WY Radiological techniques (general)
. * Applications of X-rays, gamma rays and other
. penetrating radiation, ionizing or non-ionizing.
X . X-ray techniques
XFQ . . X-ray diffraction
XFQ R . . . Bragg method (X-ray diffraction)
XFQ S . . . Laue method (X-ray diffraction)
XFQ W . . . Weissenberg method (X-ray diffraction)
XT . Gamma ray techniques
B6M Particulate radiation techniques
. * Add to B6M letters F/G following B; eg
B6M FQ . Particle diffraction
FR . Particle interferometry
GB . Particle beam techniques, bombardment
. (particles) techniques
. . * Add to B6M GB letters D/G following BM7 IM
. . * Add to B6 letters N/Q following B; eg.
B6N DC . Cosmic ray techniques
GO . Photon techniques, photonics
. . * Using photons instead of electrons.
P . Electron techniques
. . * For beta ray techniques, see nuclear radioactivity
. . techniques B6O FKR
P4 . . Instrument components
P4F V . . . Electron lenses
PFQ . . Electron diffraction
PFR . . Electron interferometry
PGB . . Electron beam techniques, electron optics
. . techniques
. . . * See also Electron optics imaging B7I NP
V . Proton techniques
W . Neutron techniques
WFQ . . Neutron diffraction
B6O . Nuclear reaction techniques
. . * For gamma ray techniques, see B6L Y.
B6O FK . . Radioactivity techniques
FKM N . . . Radioactivation analysis
. . . . * Studies artificially created nuclei in order to
. . . . elucidate the isotopes produced by
. . . . bombardment.
. . . . * For activation analysis, see Chemical analysis
. . . . C.
. . . . * For radioactive isotope techniques, see
. . . . Tracer techniques B7P.
FKR . Alpha radiation techniques
FKS . Beta ray techniques

Investigative techniques

[Operations]
[Investigative techniques B62]
[Physical methods in investigation B69]
[Mechanical techniques B6B]
. [Radiation techniques electromagnetism B6K]
. . [Particulate radiation techniques B6M]
. . . [Beta ray techniques B6OFKS]

B6O FO	. . . Nuclear resonance techniques
	. . . . * For nuclear magnetic resonance imaging, see B7I O
B6P	. . . Atom techniques
B6Q	. . . Molecular techniques
B6Q U	. . . Ion techniques
U72 N	. . . . Ion beam analysis
	. . . . *Special procedures*
U7S	. . . . . Ionic implantation
U7T	. . . . . Ion exchange techniques
	. . . . *Processes & properties*
UEY	. . . . Ion irradiation techniques
UFQ	. . . . . Ion diffraction
X	Vacuum techniques
B6R	Bulk matter techniques
	. * For sonic techniques, see B6E; for thermal techniques, see B6G P.
B6R N	. States of matter(
NR	. . Phase transition techniques, thermodynamic phases
B6T CW	. . Gas dynamics techniques
	. . . * For electrical discharge techniques, see B6I BS.
B6U CH	. . Hydrostatic techniques
CW	. . Hydrodynamic techniques
B6W	. . Crystallographic techniques
B6X	. Chemical techniques in physics
B6X CE	. . Electrochemical techniques in physics
CEL	. . . Electrolytic techniques
X	. Biological techniques in physics
B6Y	*Techniques special to a context*

[Physics B]
[Operations & agents in physics]
[Practical physics B36]
[Operations]
[Investigative techniques B62]
[Physical methods in investigation B69]
. [Techniques special to a context B6Y]

	Techniques by action on phenomenon
	. * Add to B7 numbers & letters 2,A/S following AY7; eg
B72	. Control techniques
B73	. Production techniques
B74 G	. Detecting & indicating techniques
L	. . Indicating
N	. . . Telemetry
T	. Recording techniques
V	. . Scanning techniques
B75	. Counting
B76	. Measurement
B7A	. Testing & evaluation
B7C	. . Monitoring
B7F	. Modelling & simulation
B7G	. Prediction, forecasting
B7H	. Visualizing & imaging techniques
B7H L	. . Visualizing techniques general
B7I	. . Imaging techniques
B7I O	. . . Nuclear magnetic resonance imaging
B7J	. . . Microscopy
	. . . . *By wavelength*
B7J L	. . . . Optical microscopy
LX	. . . . X-ray microscopy
M	. . . . Particle microscopy
P	. . . . . Electron microscopy
SO	. . . . Nuclear magnetic resonance microscopy
SU	. . . . Ion microscopy
B7K	. . . Holography techniques
B7K KS	. . . . Acoustic holography
T	. . . Photographic techniques in physics
B7L	. . . Radiography in physics
B7L V	. . . Tomography
B7M	. Spectroscopy, spectrography, spectrum analysis
	. . *By spectrum*
B7M KM	. . . Radiofrequency spectroscopy
KN	. . . . Nuclear magnetic resonance spectroscopy
KU	. . . Microwave spectroscopy
KX	. . . . Electron spin resonance spectroscopy
L	. . . Optical spectroscopy
LX	. . . X-ray spectroscopy
M	. . . Particulate spectra techniques
	. . . . * For mass spectroscopy, see B7O P.
NJ	. . . . Photon correlation spectroscopy

Techniques by action on phenomenon

[Physics B]
[Investigative techniques B62]
[Techniques by action on phenomenon]
. [Spectroscopy B7M]
. . [By spectrum]
. . . [Particulate spectra techniques B7MM]
. . . . [Photon correlation spectroscopy B7MNJ]

B7M NP . . . . Electron spectroscopy, electron emission spectroscopy
O . . . . Nuclear spectroscopy
P . . . . Atomic spectroscopy
QU . . . . Ion spectroscopy
. . *By various physical constants*
B7N . . . Spectrometry
. . . . * See note under Spectroscopy AY7 M; if in doubt, prefer B7M.
B7N S . . . Spectrophotometry
. . . . * Measures intensity of wavelength in optical spectra.
B7O . . . Mass spectroscopy
. . . . * Measures atomic mass by separating beam of ions into components reflecting different mass/charge ratios. Not strictly spectroscopy, since does not utilize the separation of wavelengths.
B7P . . Tracer techniques
B7P G . . . Radioactive isotope techniques
B7Q . *Techniques special to a context*
. . * For example, BWQ D7S Crystal defects - Doping.

Special forms of enquiry
B7W L . Special environments investigative techniques
B7X . . Vacuums
B7X H . . High altitude research environments
J . . Space research environments
N . Non-experimental research
B82 . Experimental physics general
. . * If distinguished from practical physics.
. . * See notes at AY3 2 regarding relations between experimental research and practical scientific work. For practical work in physics (which includes experimental methods, equipment, etc.), see B36.
. *By broad objective*
B85 . . Fundamental research
B87 . . Oriented research

[Physics B]
[Practical physics B36]
[Operations]
[Investigative techniques B62]
. [Special forms of enquiry]
. . [By broad objective]
. . . [Oriented research B87]

B8B Theoretical physics
. * For practical physics, see B36.
. *Particular theories in physics*
. . * Theories relating to a specific field, large or small, go with that field. Provision of notation at this point allows the qualification of any specific subject by its special theories; eg BGR 8TB Falling body theory of gravity.
. . * Some major theories in general physics are often (and in some cases usually) referred to for historical reasons as theories of mechanics (eg classical mechanics, quantum mechanics). The preferred arrangement is to locate them here.
. . * An alternative (not recommended) is to locate them under mechanics specifically, at BB8. In such cases, their classmark will have an additional initial B; eg quantum theory becomes BB8 M.
B8D . . Classical physics, classical mechanics, Newtonian mechanics, ensemble theory, non-quantized physics
. . . * The term 'mechanics' on its own usually assumes this.
B8D N . . . Newton's laws of mechanics
B8E . . Statistical mechanics, many-particles systems, statistical thermodynamics
. . . * Basic 19th century theories, applicable mainly to bulk matter (BR).
. . . * See also Quantum theory B8M
B8E K . . . Kinetic theory (general), molecular theory
R . . . Ergodic theory
B8F . . Field theory (general)
. . . * For unified field theory, see particle theory BM8 FG.
. . . *Properties*
B8F 97 . . . . Potential (field theory)
D . . . Classical theory
DM . . . . Maxwell theory (field theory)
G . . . Unified field theory
. . . . * For grand unified theory, see BM8 FJ.
L . . . . Supersymmetry theory
R . . . Gauge theory
S . . . . Gauge transformation
T . . . . Gauge invariance
B8G P . . Non-field theories
B8H . . Relativity theory, relativistic mechanics

Theoretical physics

[Physics B]
[Operations & agents in physics]
[Practical physics B36]
[Operations]
[Theoretical physics B8B]
[Particular theories in physics]
[Relativity theory B8H]

. * See also Gravitation BGR;
. Astronomy & astrophysics DA
. * For relativistic treatment of specific
. phenomena, see the phenomenon; eg
. relativistic scattering theory
. (elementary particles) BMF T8H.
B8H Q . Space-time continuum, four-
. dimensional continuum
R . . Interval (space-time)
S . . Events (space-time)
B8J . General relativity
. . * See also Cosmology DA
B8K . Special relativity
B8M Quantum theory, quantum
mechanics
. * Further details are given at BM8 M
. Particle physics. Details from there
. may be used here if necessary.
. Examples are given below.
. * For quantum theory of specific
. phenomena, see the phenomenon; eg
. quantum electrodynamics BNG 8M;
. quantum chromodynamics BNR 8M.
B8M 27 . History
27E . . Early period
27F . . . Planck quantum theory
27H . . Later period
. . . * 1924 onwards.
27J . . . Einstein quantum theory
27K . . . Bohr quantum theory
. Statistics
2X . . Quantum statistics
. . . * See also Elementary particles -
. . . Energy levels BMB D
2XY AB . . . Bose-Einstein statistics
2XY AF . . . Fermi-Dirac statistics
. Constants
76C . . Planck constant
. Particular quantum theories
D . . Classical theory
DT . . . Semi-classical quantum
. . . theory
DV . . . Non-relativistic quantum
. . . theory
. . Field theory
. . . * For Gauge field theory, see
. . . BM8 MFR.
F . . . Quantum field theory,
. . . quantified fields,
. . . quantized fields

[Operations]
[Theoretical physics B8B]
[Particular theories in physics]
[Quantum theory B8M]
[Particular quantum theories]
. [Field theory]
. . [Quantum field theory B8MF]

. . . * For quantum field theory of specific
. . . phenomena, see the phenomenon - e.g.
. . . meson field theory BNS 8MF.
B8M FN . . . Non-linear field theory
FX . . . Axiomatic field theory
H . Relativistic quantum theory
HF . . Relativistic quantum field theory
Quantum properties
N . Quantization
P . Complementarity
R . Quantum number (theory)
. . * For quantum number properties, see
. . Particles BMM D.
RW . . Correspondence principle
S . . Pauli exclusion principle
. . . * See also Periodic table CH; Spin
. . . (particles) BMM K
T . . Principal quantum number
B8N *Special quantum properties, quantum numbers*
. * This allows the qualification of a particular
. property if its quantization is the subject.
B8O *Special quantum effects*
. * For example, BRQ BBQ 8O Surfaces -
. Potential energy - Potential barrier -
. Tunnellling.
B8O V Two-state quantum theory
B8P Wave mechanics (quantum theory),
corpuscular waves (quantum theory)
B8P 2M . Mathematics
2M8 GE . . Perturbation theory (quantum
. . mechanics)
2M8 L . . Wave functions
. . . Named functions
2M8 NH . . . . Hamiltonian functions,
. . . . Hamiltonians
. . . Named equations
2M9 NS . . . . Schrodinger wave equation
2TB . Matrices
. Particular theories
QU . . Bohr theory (wave mechanics)
S . . Schrodinger wave mechanics
U . . Uncertainty principle, indeterminacy
. . principle, Heisenberg principle
B8R B Matrix mechanics

[Physics B]
[Operations & agents in physics]
. . [Operations]
. . . [Theoretical physics B8B]
. . . . [Particular theories in physics]
. . . . . [Quantum theory B8M]
. . . . . . [Matrix mechanics B8RB]

B8R B8V . . . . Heisenberg representation
General processes/properties in physics
. * Classes B92/99 are for use only as qualifiers. See
. explanatory note at AY9.
. * Add to B9 numbers 2/9 following AY9 and letters
. G/S following AY (with the additions indicated); eg
B92 D . Distribution
E . . Incidence
F . Invariability, constancy
. . * For constants, see Measurement B76 C
H . Variation, change
J . . Rate of change, gradient
K . . Decrease
L . . Increase
N . . Cyclical change
S . Development
T . . Formation, origin
V . . Growth
B94 . Conditions, parameters, environments,
. influences
B94 BC . . Degrees of freedom
C . . Critical point, critical state
E . . Volume conditions
F . . . Constant volume
G . . . Decreasing volume
H . . . Increasing volume
J . . Pressure conditions
JC . . . Critical pressure
K . . . Constant pressure conditions, isobaric
. . . conditions
L . . . Decreasing pressure conditions
M . . . Increasing pressure conditions
N . . . Other pressure conditions
O . . Velocity conditions
P . . Thermal conditions
PC . . . Critical temperature
Q . . . Constant temperature, isothermal
. . . conditions
R . . . Adiabatic conditions
S . . . Decreasing temperature conditions
V . . . Increasing temperature conditions
Y . . Electrical & magnetic field conditions
B95 . . . Electrical field conditions
B96 . . . Magnetic field conditions
B97 . . . Radiation conditions
B98 . . Other conditions
Special properties
B9B . Physical dimensions, dimensions of a
. physical quantity
. . * For mass, see BCJ; for measurement, see B76.

[Physics B]
[Special properties]
[Physical dimensions B9B]

B9B 2X Dimensional analysis
Processes
92K . Contraction
92L . Expansion
Dimensional properties
G . Degree of dimension, number
. . * Use for any qualification or degree of a dimension
. . - e.g. double, very, optimal.
H . Nodes, points (dimension)
I . Indexes
IV . Dimensionless groups
J . Similarity
K . . Dynamic similarity, similarity principle
. . . * See also Similarity parameters (fluid flow)
. . . BSB 9BK
Q . Frames of reference
R . . Inertial reference frame
. *By number of dimensions*
T . . One-dimensional
U . . Two-dimensional
V . . Three-dimensional
W . . Multi-dimensional
. . . * More than three dimensions.
B9C Time
. * The dimension of the physical universe which at a
. given place orders the sequence of events.
. * See also Space-time continuum B8H Q
. Measurement
B9C 76 . . Horology (physics)
. . *Processes*
92K . . . Contraction time dimension
92L . . . Expansion time dimension
9FC . . . Reversibility (time)
. . . . * See also Relativity theory B8H
9FE . . . Irreversibility (time)
9FG . . . . Arrow of time
C . Time conditions
EN . . Ante, prior
EP . . Post (time)
F . . Rate (time)
G . . . Slow
H . . . Fast
I . . Duration, life
. . . * For lifetime as a quantum property, BMM FH
J . . . Momentary, transient
K . . . Short duration
L . . . Long duration
M . . . Mean life
N . . . Half life, lifetime
P . . Time intervals
Q . . . Frequency (time dimension)
QS . . . Short intervals
QU . . . Long intervals
R . . Instantaneous, immediate

[Physics B]
[Special properties]
[Physical dimensions B9B]
[Time B9C]
. [Time conditions B9CC]
. . [Instantaneous B9CR]

B9C S . . Continuous (time)
T . . Discontinuous (time), discrete (time)
U . . Periodic (time conditions)
V . . . Isochronous
B9D Space, spatial dimension, size
. * See also Space-time continuum B8H Q; Shape, configuration B9J V
. *Processes*
B9D 92K . . Contraction (space dimensions)
92L . . Expansion (space dimensions)
. *Constituents*
D . . Aether, ether
. . . * Hypothetical medium once thought to fill all space.
E . Direction
EG . . Orientation
EK . . Isotropic
EL . . Anisotropic
EP . Position, location
. . * For example, back, front, side, end.
F . One-dimensional spaces, lines, linear dimensions
G . . Distance
H . . . Mean free path
I . . Length
J . . Width, breadth
JN . . Height
K . . Radius
L . . Diameter
M . . . Chord (dimensions)
N . . Perimeter, circumference
P . Two-dimensional spaces, planes
PQ . . Angles
PR . . Area
PS . . . Sectors (space dimensions)
PU . . Surfaces (dimensions)
Q . Three-dimensional spaces, bodies (3-d space)
R . . Cross section (3-d space)
RT . . Edge conditions
S . . Volume
S94 C . . . Critical volume
SP . . Solid bodies
SQ . . Hollow bodies
SS . . Homogeneous bodies
ST . . Non-homogeneous bodies, heterogeneous bodies
T . . Physical field
. . . * For fields of force, use BBH.
. . . * See also Field theory B8F
. Multidimensional

[Physics B]
[Special properties]
[Physical dimensions B9B]
. [Space B9D]
. . [Multidimensional]

B9D V . . . Hyperspace, multidimensional space
. . . . * More than 3 dimensions.
. . . . * For space-time (3-d space and time), see Relativity B8H Q.
X . . . Non-Euclidean space
B9G Systems characteristics
B9G C . Systems behaviour
E . . Disturbance, perturbation
G . . . Temporary disturbances, transients
P . . State of system
T . . . Constancy, invariance
V . . . . Conservation laws
. . . . . * For specific laws, see thing conserved; eg conservation of mass & energy BAE 9GV.
. . . . . * See also theory of elementary particles BNB 8B
B9H . . . Variability
B9H H . . . . Change of state of system
J . . . . . Periodic change
. . . . . . * For periodic motion, see BDS.
K . . . . . . Cycles periodic change
N . . . . . Response systems
Q . . . . . . Hysteresis
R . . . . . Autonomous events
U . . . Stability of systems
V . . . . Homeostasis, self-regulation
W . . . Instability of systems
B9I H . . . Adaptive behaviour
B9J . Structure of systems physics
B9J S . . Symmetry, homogeneity
. . . * For symmetry reflecting invariance principle in particles, see BMM B
T . . . Parity (general), space reflection symmetry, mirror symmetry
T9G V . . . . Conservation
. . . . . * For P invariance (parity conservation), see BMM E9G V.
V . . Shape, configuration
W . Networks systems
Types of systems
B9K V . Continuous systems
X . Discontinuous systems, discrete systems
. . * See also Quantum mechanics B8M
B9L R . Linear systems
B9M . Non-linear systems
B9N P . Adaptive systems

Energy interactions & forms

[Physics B]
[Special properties]
. [Types of systems]
. . [Adaptive systems B9NP]

BAE Energy & matter (together)
. * Energy is the capacity for doing work.
. Interconversion between its different forms (potential,
. kinetic, electrical, chemical, etc) can occur only in the
. presence of matter. Energy can only exist in the
. absence of matter when it is in the form of radiant
. energy.
. * Matter is a specialized form of energy having the
. attributes of mass and of extension in space and time.
. For general works on matter per se, see BLY.
. * For Mechanical energy, see BBB M; for Matter, see
. BLY.
. Conservation
BAE 9GV . . Conservation of mass & energy
9GW . Equivalence of energy & matter

BAF Energy interactions & forms
BAG . Thermodynamics
. . * Interactions between energy systems and their
. . effect on the states of those systems.
. . * For thermal phenomena and heat in general, see
. . Bulk matter physics BRG P.
. . * For statistical theory of thermodynamics see B8E.
. . Particular theories
BAG 8D . . . Classical thermodynamics, ideal
. . . processes, reversible processes
. . . (classical theory)
. . . . * For reversible thermodynamics, see BAP D.
8DP . . . Equilibrium thermodynamics
92N . . . Cycles
. . . . * See BAP Y
9G . . Systems characteristics
9IC . . . Reversible processes
. . . . * See BAP D
9IE . . . Irreversible processes
. . . . * See BAP E
BAH . . Principles of thermodynamics, laws of
. . thermodynamics
BAI . . . Zeroth law of thermodynamics
BAJ . . . First law of thermodynamics,
. . . conservation of energy
BAK . . . . Enthalpy
BAL . . . Second law of thermodynamics
BAM . . . . Entropy
. . . . . * See also irreversible thermodynamics BAP E
BAM J . . . . . Entropy of substances
BAN . . . Third law of thermodynamics, Nernst
. . . heat theorem
. . *Thermodynamic processes & properties*
. . . * For internal energy, see BBB V.

[Physics B]
[Energy interactions & forms BAF]
[Thermodynamics BAG]
[Thermodynamic processes & properties]

* See also Thermal properties (bulk matter)
BRG P
BAO Fluctuation phenomena, variables
(thermodynamics)
BAP D Reversible thermodynamics, reversible
processes
E Irreversible thermodynamics, non-
equilibrium thermodynamics
. * Extension of classical theory to cover real
. dynamical processes; eg in biology.
. * See also Entropy BAM
H Critical state, critical point phenomena
N Free energy, thermodynamic potential,
Gibbs function
R Adiabatic processes
T Isothermal processes
V Isentropic processes
W Polytropic processes
Y Cycles (thermodynamics)
BAQ . *Named cycles*
. . * Arrange A/Z; eg
BAQ C . . Carnot cycle
BAT Transport processes (general)
. * Transfer of mass, momentum or energy in a
. system.
. * For transport processes & properties in
. particular energy systems, see the system - e.g.
. electrical conductivity, neutron transport,
. dielectric relaxation, viscosity (fluids).
. Theory
BAT 8EK . . Kinetic theory
92N . Cycles (transport processes)
BAU . Scattering (transport processes)
BAU D . Dissipation (transport processes)
BAV . Diffusion (transport processes)
BAW . Viscosity
. . * See Fluids BSA W
BAX . Mass transfer
. . * See also Heat transfer BRG Q

[Physics B]
[Energy interactions & forms BAF]
[Thermodynamics BAG]
. [Thermodynamic processes & properties]
. . [Transport processes general BAT]
. . . [Mass transfer BAX]

BB Mechanics
. * Study of the behaviour of physical systems under
. the action of forces, especially with the motion and
. equilibrium of bodies in a particular frame of
. reference. Sometimes used with wider meaning, to
. include the behaviour of all physical systems under
. all interactions, when it is barely distinguishable
. from physics as a whole.
BB2 M8L . Mathematical functions
M8N H . . Hamiltonian functions
M9L . Equations
M9N H . . Hamiltonian equations
M9N L . . Lagrangian equations
. *Named systems*
P2H . . Hamiltonian systems
P2L . . Lagrangian mechanics
S . . Geometric mechanics
BB8 B . Theory
. . Particular theories
. . . * Classical theory, relativity theory, quantum
. . . theory, etc are treated as theories of general
. . . physics (see B8D/Y).
. . . * An alternative (not recommended) is to
. . . subordinate them here,to mechanics. If this
. . . option is taken:
. . . * Add to BB8 letters D/Y following B8; eg
D . . Classical mechanics
. Continuous systems
BB9 KV . . Continuous mechanics, classical
. . mechanics of continuous matter,
. . continuous media (mechanics)
KVK . . . Mechanical contact
. Discrete systems
KX . . Classical mechanics of discrete systems
KXK . . . Few-body theory
. . . . * See also Gravitation BGR
KXM . . . . Two-body problem
KXP . . . . Three-body problem
KXR . . . N-body problem, many-body problem
BBB . Energy
. . * Capacity of a body or system for doing work.
. . * For activation energy, see Chemistry C; for energy
. . levels and energy bands, see Elementary particles
. . BMB D.
BBB 92D . . Distribution of energy
9GV . . Conservation of energy
. . . * See Thermodynamics BAJ
M . . Mechanical energy
. . . * A measure of a system's capacity to do work.
P . . . Potential energy, energy of space,
. . . energy of position

[Physics B]
[Energy interactions & forms BAF]
[Mechanics BB]
[Energy BBB]
. [Mechanical energy BBBM]
. . [Potential energy BBBP]

BBB T . . Kinetic energy, energy of motion
. . . * For motion, see BCS.
. . . * See also friction BVQ CA
V . . Internal energy, thermodynamic energy
. . . * Sum of the potential and kinetic energies of
. . . molecular interactions.
BBC . . Energy ranges
. . . * See particle physics, BMB C
BBD . . Energy levels
. . . * See particle physics, BMB D/E
BBF . . Energy bands
. . . * See condensed matter BTX BF
BBF N . *Energy concepts special to a subject*
. . * Eg Nuclear collisions - Excitation energy
. . BOF SBF N.
BBG Force
. * Any action which alters or tends to alter a body's
. state of rest or of uniform motion in a straight line.
. * For an expansion of this general class, see nuclear
. forces BOB G. For Gravity, see BGR.
BBH . Fields of force
. . * A 3-dimensional space throughout which forces
. . can act and in which energy is available.
BBH J . . Gradient (fields)
K . . Strength of field
L . . Interaction (field)
O . . Field effects
Q . . Lines of force, force field
Q2U . . . Graphs
Q2V . . . . Force polygons
Q2W . . . . Catenaries
BBI G . Work
. . * Result of the exchange of energy between field
. . and body experiencing forces.
H . . Power
. . . * Rate of doing work.
HM . . . Turning moment
J . . Mechanical efficiency
JM . . . Velocity ratio, mechanical advantage
JS . . Virtual work
M . Moment, torque, turning effect
. . * The moment of a force about a point is the
. . product of the force and the perpendicular
. . distance of its line of action from the point or axis
. . of rotation.
. . * For moment of momentum, see Angular
. . momentum BDN CV.
N . . Centrifugal moment
P . . Moment of inertia
. *Types of forces*
Y . . Generalized force

[Physics B]
[Energy interactions & forms BAF]
[Mechanics BB]
[Force BBG]
[Types of forces]
[Generalized force BBIY]

. * Quotient of all forces acting in a system.

BBJ Pressure
. * Force acting per unit area.
. * For compressibility, see Elasticity BCB; for load, see BBQ.
BBJ 92D . Distribution
94 . . Pressure conditions
94E . . . Critical pressure
94K . . . Constant pressure, isobaric conditions
. Properties
BH . . Pressure fields
BHJ . . . Pressure gradients
BHK . . . . Pressure gradient force
MP . . Pressure effect(
NN . . . Centre of pressure
NP . . . Compression
NQ . . . Head of pressure
. . . . * See BSB JO (fluids)
. *Types of pressure*
P . . Low pressure
Q . . High pressure
. . . * For high pressure physics, see BRB JQ
S . . Other
. . . * Add to BBJ S letters B/N following BSB JS if applicable; eg BBJ SF Induced pressure.
TC Conservative force
TD Restoring force
TE Elastic force
. * See Elasticity BCB
TI Impulsive force
U . Impact, collision of bodies
. . * Physical effect of impulsive force.
. . * See also Ballistics BGR G
VP Propulsion, thrust, push (propulsion)
VR Traction, pull
VS Central force, Single force
VU Multiple force
W External forces
. * Transmitted throughout a body
WK . Impressed force, action (mechanics)
WL . Reaction (mechanics)
WM . Body forces
WN . Surface forces
WP . Integral forces
. . * Between neighbouring parts of a body.

[Physics B]
[Energy interactions & forms BAF]
[Mechanics BB]
[Force BBG]
[Types of forces]
[External forces BBJW]
[Integral forces BBJWP]

BBK Deforming forces, deformation, distortion
. * The detailed schedule for this class is given in bulk matter physics (at solid state BVB K), to which it is largely applicable.
. * Add to B letters BK/CD following BV (Solids) so far as applicable; eg
BBK L . Stress-strain relationships
N . . Limit of proportionality
P . . Strength (stress-strain)
Q . . . Yield point
BBL . Stress (general)
. . * Deforming force per metre.
. . * For Creep, see Bulk matter BVB YK.
BBL V . . Stress components
. . . * Internal forces between contiguous parts of a body.
. . . * For normal stress, see BBO.
X . . . Tension
. . *Types of stresses*
BBM T . . . Bending stress
BBO . . . Normal stress, direct stress
BBO T . . . . Tensile stress, tensional stress
. . . . . * For compression, see BBJ NP; for compressibility, see BSB QS.
BBP . . . Shear stress
BBQ . . Loading, load
. . . * Force, external to a machine, which the work output must overcome.
. . . * The detailed schedule for this class is given under Gases (bulk matter) BTB Q.
. . . * For load as weight, see BCK; see also Stress BBL.
BBQ S . . . Compressibility
BBT . Strain
. . * Deformation resulting from stress.
BBU N . . Torsion
BBV E . . Elastic deformation
BBW P . . Plastic deformation
BBY D . . Failure (strain)

[Energy interactions & forms BAF]
[Mechanics BB]
[Force BBG]
. [Types of forces]
. . [External forces BBJW]
. . . [Deforming forces BBK]
. . . . [Strain BBT]
. . . . . [Failure strain BBYD]

BCB . . . . Elasticity
. . . . . * Property of a physical system allowing it to
. . . . . return to the original physical state after
. . . . . removal of a stress.
BCB E . . . . . Modulus of elasticity, elastic
. . . . . constants
. . . . . . * Ratio of stress to strain.
BCC C . . . . . Viscoelasticity
BCD . . Internal forces
BCE . . Inertial forces, effective forces, kinetic
. . reactions
. . . * For inertia, see BCI
. . . * See also Dynamic equilibrium (statics)
. . . BCO V; Acceleration BDD
BCE G . . . Centrifugal force
H . . . . Coriolis force
J . . . . Centripetal force
L . . Couple, coupling
. . . * Two equal & opposite parallel forces acting on
. . . body.
N . . Attraction, attractive force, potential
. . (attraction)
R . . Repulsion
BCF . *Forces special to a context*
BCH Statics
. * Behaviour of bodies at rest relative to a given
. frame of reference; the forces acting on them cancel
. each other out and produce a state of equilibrium in
. which the bodies are stationary or moving with
. constant velocity; i.e. acceleration and torque are
. zero.
. * See also Hydrostatics BUC H ; Torque BBI M
BCH 2X5 8C . Graphical methods
2XS . . Analytical statics
. Moments
BIM . . Static moment
BCI . Inertia
. . * Ability of a body to resist changes in its state of
. . rest or of uniform motion in a straight line.
. . * For rotational inertia, see rotary motion
. . BDN CI.
. . * See also motion BCS; Hysteresis (general) B9H Q
BCI BIM . . Moments of inertia
BIN . . Angular inertia
BIP . . Products of inertia
BCJ . Mass
. . Relativity theory
BCJ 8H . . . Relativistic mass
BCK . Weight, load (weight)

[Physics B]
[Energy interactions & forms BAF]
[Mechanics BB]
[Statics BCH]
. [Weight BCK]

. . * See also Gravity BGR
BCL . Density, API gravity, bulk density
BCL L . . Specific gravity, relative density
P . . Low density
T . . High density
W . . Wet density
BCM . Composition & resolution of forces
. . * See also Stress BBL; Couple BCE L
BCN . . Equilibrium
. . . * State of a body at rest or moving with
. . . constant velocity.
. . . * For Phase equilibrium, see States of matter
. . . BRN T.
BCN P . . . Balancing
T . . . Parallelogram of forces, polygon of
. . . forces
U . . . Parallel forces
V . . . Non-parallel forces
W . . . Quasi-equilibrium
X . . . Non-vanishing equilibrium
Y . . . Neutral equilibrium, indifferent
. . . equilibrium
BCO . . . Stable equilibrium, static equilibrium
BCO P . . . . Least energy principle
Q . . . . Neutral equilibrium
U . . . Unstable equilibrium
V . . . Dynamic equilibrium
BCP . . Stability
. . . * General property of systems whereby the
. . . system returns to state of equilibrium after
. . . disturbance.
BCP Q . . . Absolute stability
R . . . Asymptotic stability
S . . . Static stability
BCQ . . . Dynamic stability
BCR . . Instability
. . . * For balancing, see BCN P
BCR S . . . Dynamic instability
U . Suspending
V . Equalizing
BCS Motion
. * A continuous change of position of a body.
BCS 8DN . Newton's laws of motion
T . Hamilton's principle (motion)
U . D'Alembert's principle
V . Principle of least constraint, constrained
. motion
W . Principle of least action
W2M 9 . . Equations
W2M 9N . . . Named equations
. . . . * Arrange A/Z; eg
W2M 9NL . . . . Lagrange's equations of motion

Dynamics

[Energy interactions & forms BAF]
[Mechanics BB]
[Motion BCS]
. [Principle of least action BCSW]
. . [Equations BCSW2M9]
. . . [Named equations BCSW2M9N]
. . . . [Lagrange's equations of motion BCSW2M9NL]

. *Properties & processes*
BCT . . Immittance (mechanics)
BCT S . . . Mechanical admittance
. . . . * Reciprocal of impedance.
T . . . Mechanical impedance
. . . . * Ratio of driving force to response.
U . . . . Mechanical resistance
V . . . . Mechanical reactance
BCU . Momentum, linear momentum, vector momentum
. . * Quantity of motion of a body; product of mass and velocity.
. . * For angular momentum, see Rotation BDN CV.
BCU Q . . Impulse (momentum)
BCV . . Angular momentum
BCW . Flux
. . * A measure of the strength of a field of force (eg a rate of flow) through a given area.
BCX Dynamics, force & motion
. * Deals with the forces which change or produce the motions of bodies.
. *Properties*
BCX 9BQ . . Frames of reference
9BS . . . Rotational frames of reference
9M . . Non-linear dynamics
E . Initial value problem, transient problem
. . * Value which determines subsequent state of a system.
G . Steady state problem
. . * State of the system remains unchanged in time after all the transients resulting from changes have been removed.
BDA . Kinematics, pure motion
. . * Motion independent of considerations of mass or force; geometry of motion.
. . * See also Kinetic theory (general) BAT 8ES
BDA S . . Speed
. . . * A scalar quantity; use only if this is significant. Otherwise, use Velocity BDC.
BDB . . Displacement
. . . * Change of position.
. . . Principles
BDB E . . . . Laws of virtual displacement
BDC . . Velocity, speed (velocity)
BDC 92D . . . Velocity distribution, velocity gradient
E . . . Principle of virtual velocity
G . . . Linear velocity
H . . . Angular velocity
J . . . Relative velocity
. . . . * See also Doppler effect BFA DCK

[Physics B]
[Energy interactions & forms BAF]
[Mechanics BB]
[Dynamics BCX]
[Kinematics BDA]
[Velocity BDC]
. [Relative velocity BDCJ]

BDC M . Terminal velocity
N . Low speed
O . Medium speed
P . High speed
BDD Acceleration, change of motion
. * See also Force BBG
BDD F . Increase in speed
. . * 'Acceleration' in the popular sense.
H . Deceleration
J . Linear acceleration
L . Angular acceleration
. . * See also Rotation BDN
N . Coriolis acceleration
BDE Kinetics
. * Effect of forces or torques on motion.
. * Definition of this term varies; sometimes it is equated with kinematics, sometimes with dynamics.
. * For kinetics as rate of chemical reaction, see C; for kinetic theory (general), see BAT 8ES; for kinetic theory of fluids, see BSD E8B.
Forms of motion
. * Many of these relate almost entirely to particular states of matter (e.g. fluid flow). Classes are expanded where necessary for very specialized forms; e.g. attitudes in aerodynamics.
. * For Acceleration, see BDD; for acceleration due to gravity, see Gravitation BGR DD.
. *By entities in motion*
BDK B . . Motion of points
C . . Motion of extended figures
D . . . Rigid figures (motion), rigid bodies (motion), solid bodies (motion)
E . . . Deformable figures (motion)
G . . . One-dimensional figures (motion)
L . . . Two-dimensional figures (motion), moving planes
LKH . . . . Rigid
LKJ . . . . Deformable
M . . . Three-dimensional figures (motion)
MKH . . . . Rigid
MKJ . . . . Deformable

Dynamics

[Dynamics BCX]
[Kinematics BDA]
[Forms of motion]
[By entities in motion]
. [Motion of extended figures BDKC]
. . [Three-dimensional figures motion BDKM]
. . . [Deformable BDKMKJ]

BDK Q . . . Articulated systems (motion)
. . . . * 3-d. deformable figures with rigid elements.
T . Collective motions
. . * See particle physics BNU DKT
BDL A Pressure-affected motion
. * For fluid flow, see Fluids BSB.
C Linear motion
E One-dimensional motion
L Relative motion
N Angular motion
R Rectilinear motion
. * See also Rays BGC
T Translation (motion), sliding motion
. * For friction, see Solids - Surfaces BVQ CA.
U Curvilinear motion
V . Ballistic motion
. . * For ballistic motion as a property of gravitation,
. . see BGR G.
BDM Circular motion, circulatory motion
BDM U . Uniform circular motion
W . Orbits (general)
BDN . Rotation, gyration, pre-rotation,
. revolution, revolving, rotatory motion
. . * For internal rotation, see molecular motion
. . BQB DQM.
. . Inertia
BDN CI . . . Rotary inertia
CV . . Angular momentum
CV . . . Moment of momentum
CX . . . . Complex angular momentum plane
. . *Properties*
P . . . Radius of gyration
Q . . . Euler angles
S . . Spinning, spin
T . . . Nutation
. . . . * Periodic variation in spin.
U . . . Precession
W . . Whirl
WP . . Prewhirl
BDP . Vortices, vortex motion
. . * Usually implies bulk matter.
. . * Add to BD letters P/Q following BSD so far as
. . applicable; eg
BDQ R . . Three-dimensional vortex motion
BDR Irrotational motion
. * Special to fluids.
BDR S Planetary motion
T Coplanar motion

[Physics B]
[Energy interactions & forms BAF]
[Mechanics BB]
[Dynamics BCX]
[Kinematics BDA]
[Forms of motion]
[Coplanar motion BDRT]

BDS Periodic motion, harmonic motion
. * Repetitive, periodic change in
. displacement with respect to a
. reference point.
. * There is relatively little literature on
. specific properties and processes
. treated completely generally (i.e. in
. mechanical vibration, radiation and
. wave phenomena, etc. together). So
. the detailed enumeration of properties
. and processes is given at BF Wave
. motion (much the biggest class in
. terms of literary warrant) and this
. may then be drawn on for
. synthesizing classes in the other
. contexts, as instructed below at BDU,
. BDV, BE and BF.
. * Add to BDS letters A/D following B;
. * Add to BDT letters B/W following
. BF;
. * Add to BDT X letters A/G
. following BG so far as applicable; eg
BDS 92N . Cycles (periodic motion)
BB . Energy
. . * For radiant energy in general, see
. . Radiation BEY.
BDT B . Harmonics
. . Generation
B73 D . . . Harmonic generation
. . *Elements*
BH . . . Nodes
BJ . . . Antinodes
. . *Types*
BM . . . Simple harmonic motion, SHM
. . . . * The standard component of all
. . . . vibrations.
BN . . . . Damped simple harmonic
. . . . motion
BQ . . . Anharmonics
BR . Isochronism, regular periodicity
C . Transmission
BDU . Oscillation (general)
. . * See note at BE. Use only if
. . distinguished from vibration; if in
. . doubt, prefer the joint heading at
. . BE.
. . * Add to BDU letters A/V following
. . BE; eg
BDU B . . Harmonics
C . . Transmission

[Mechanics BB]
[Dynamics BCX]
[Kinematics BDA]
[Forms of motion]
[Periodic motion BDS]
[Oscillation general BDU]
. [Transmission BDUC]

BDU VVE V	. Large amplitude oscillations
WD	. Torsional oscillations
BDV	Vibration (general)
	. * See note at BE. Use only if distinguished from oscillation; if in doubt, prefer the joint heading at BE.
	. * For mechanical oscillation, see Bulk matter BRE.
	. * Add to BDV letters A/V following BE; eg
BDV B	. Harmonics
C	. Transmission
VVE V	. Large amplitude vibration
WD	. Torsional vibrations
BE	Vibration & oscillation
	. * Usage varies: in much of the literature oscillation and vibration are treated as synonymous; often, oscillation is used in relation to the wave motion of radiation whereas the term vibration is used for the same phenomenon in bulk matter (see BRE). See notes at BDU and BDV above.
	. * Add to BE letters A/VV following BF; eg
BEB	. Harmonics
BEC	. Transmission, propagation
BED	. Frequency
BED D	. . Mode
BEP Y	. Interactions
BER U	. . Coupling
	. *Types of vibration & oscillation*
BEV V	. . *By property*
	. . . * Add to BEV V letters D/V following BFV V, so far as applicable; eg
VEV	. . . Large amplitude vibration & oscillation
W	. . *By directional & transient factors, etc.*
	. . . * See also BRE W (vibration in bulk matter) for amplification of some of these types of vibrations.
	. . . * Add to BEV W letters C/V following BFW so far as applicable;
	. . . * Add to BEV X letters A/Q following BFX so far as applicable;
	. . . * Add to BEW letters B/X following BFY so far as applicable;
	. . . * Add to BEW Y letters A/E following BG if applicable. applicable; eg
WC	. . . Isotropic oscillations
X	. . . Pulse oscillations

[Dynamics BCX]
[Forms of motion]
[Periodic motion BDS]
[Vibration & oscillation BE]
. [Types of vibration & oscillation]
. . [By directional & transient factors BEVW]
. . . [Pulse oscillations BEVX]

BEW D	. . . Torsional vibrations
J	. . . Linear oscillations
L	. . . Non-linear oscillations
R	. . . Transverse vibrations (general)
S	. . . . Shear vibrations
U	. . . Longitudinal vibrations (general)
V	. . . . Axial vibrations
	. . *By degrees of freedom*
BEX C	. . . One degree of freedom (vibration)
D	. . . Two degrees of freedom (vibrations)
E	. . . Three or more degrees of freedom
	. . *By origin as to internal/external*
G	. . . Free vibration, natural frequency vibration
H	. . . Forced vibration
	. . . . * For resonance, see BFO.
J	. . . . Transients (forced vibration)
	. . . . . * Frequency is determined by the natural frequency of the system.
	. . . . . * For transients as transmission disturbances, see BFC F.
K	. . . . Steady component
L	. . *By origin as to action*
	. . . * For example, by vibrating body.
	. . . * See Bulk matter BRE XL
BEY	Radiation (general)
	. * Energy from vibration propagated as rays, waves or a stream of particles. Often treated as synonymous with electromagnetic wave propagation.
	. * Use this position only when a distinction is drawn between radiation and wave motion; also, in particular, for general works on the special energy features of radiation. If in doubt, prefer BF Waves.
	. * For types of radiation defined by a specific energy form, see the latter; eg electromagnetic waves BK; sound waves BRG H.
	. * For ionizing radiation, see BKJ.
	. Energy
BEY BB	. . Radiant energy
BB9 2D	. . . Distribution
BCW	. . . Radiant flux, radiant power, radiant energy flux
	. . . . * See also Opacity BRL FIL; Luminous flux BRL LBB F
BCW CL	. . . . Radiant flux density, irradiance, radiant exitance, power density

Waves

[Dynamics BCX]
[Periodic motion BDS]
[Radiation general BEY]
. [Energy]
. . [Radiant energy BEYBB]
. . . [Radiant flux BEYBCW]
. . . . [Radiant flux density BEYBCWCL]

BEY BCW P . . . . Radiation intensity
. . . . . * For luminous intensity, see BRL LBB G.
BCW Q . . . . . Radiance (intensity)
. . . . . . * Radiation intensity in a specified
. . . . . . direction.
F . *Processes & properties*
. . * Add to BEY F letters B/V folowing BF so far
. . as applicable; eg
FJ . . Decay
. *Types of radiation*
. . * Add to BEY letters VV/W following BE so far
. . as applicable; eg
. . *By property*
VVP . . . Polarized radiation
. . *By direction, etc.*
VWC . . . Isotropic
WJ . . . Linear

BF Waves, wave motion
. * Add to BFA letters A/DR following B.
. * Classes BFB/BFF are qualified retroactively as
. follows (where the hyphen represents the classmark
. qualified):
. * Add to -A letters A/DR following B;
. * Add to - letters B/F following BF; eg Resonance -
. Frequency BFO D;
. * Classes BFG/BFV are qualified retroactively as
. follows (where G is used as an intercalator):
. * Add to -A letters A/DR following B;
. * Add to - letters B/F following BF;
. * Add to -G letters G/Y following BF; eg
. Collisions - Absorption BFS GL.
. *Operations*
BF7 2 . . Control wave motion
2D . . . Damping
. . . . * For damping as a natural process, see Decay
. . . . BFC N.
3D . . Generation of waves
3K . . . *By source, medium*
. . . . * For example, Sound- -Vibrating bodies
. . . . BRG H73 LB.
. *Properties*
BF9 DI . . Wavelength
. . . * See also Frequency BFD
. Energy
BFA BB . . Wave energy, radiant energy (waves),
. . radiative power (waves)
. . . * Add to BFA BB letters F/J following
. . . BEY BB; eg
BBF . . . Radiant power

[Kinematics BDA]
[Forms of motion]
[Periodic motion BDS]
[Waves BF]
[Energy]
. [Wave energy BFABB]
. . [Radiant power BFABBF]

. . . Radiant flux density
BFA BBG . . . . Magnitude (waves), power density
. . . . (waves)
Force
BH . Field
BHK . . Field strength
Stability
BLR . Relaxation (wave stability)
Kinematics
DA . Relative velocity
DCJ . . Doppler effect
Properties & processes
BFB . Harmonics
. . * Add to BFB letters A/R following BDT B;
. . eg
BFB H . . Nodes
BFC . Transmission (radiation), propagation
. (radiation)
. . Coefficients
BFC 2QK . . . Transmission coefficient
. . . . * Reciprocal of opacity (see Optics
. . . . BRL FIL).
. . . . *For transmittance, see BFC H.
BHK . . Field strength
F . . Transients (transmission)
. . . * Temporary disturbances. For transients as
. . . types of vibration, see BEX J.
FR . . Propagation anomaly
. . . * See also Resonance BFO
FS . . . Spreading anomaly, spreading loss
. . . (waves)
G . . Propagation loss
. . . * See also Attenuation loss BFC MG;
. . . Scattering loss BFT CG
H . . Transmittance
I . . Transmittivity
K . . Modulation
. . . * Changes in various parameters.
L . . Amplification, gain
LP . . . High gain
LQ . . . Low gain
M . . Attenuation
. . . * Reduction of a radiation quantity.
MG . . . Attenuation loss, dissipation
N . . . Damping (waves)
P . . Transmission modes
PU . . . Multimodes
. . *Types of propagation*
Q . . . Unguided propagation
R . . . Guided propagation

[Forms of motion]
[Periodic motion BDS]
[Waves BF]
[Properties & processes]
[Transmission radiation BFC]
. [Types of propagation]
. . [Guided propagation BFCR]

BFC RR . . . Network-guided propagation
RS . . . Conductive lines guided propagation
RV . . . Non-conductive lines guided
. . . propagation
RX . . . Hollow conductor guided propagation
S . . *Special to a context*
. . . * For example, Solar propagation BKM CS;
. . . Baryon photoproduction BNT FCS.
BFD Frequency
. * See also Electro-magnetic waves by frequency
. BKL/BL; Wavelength BF9 DI
BFD D . Mode
DS . Frequency shift
E . Frequency bands, frequency ranges
E9D J . . Band width
EP . . Basebands
ER . . Broad bands
ET . . Narrow bands
F . Cut-off frequency
G . Critical frequency
HB . Variable frequency
HD . Difference frequency
HL . Double frequency
HS . Single frequency
L . Low frequency
N . Medium frequency
O . High frequency
P . . Very high frequency
. . . * For short waves, see BKO.
BFE . Spectra (general)
. . * For spectroscopy, see Physical methods B7M.
. . * For spectra of particular processes, see
. . process; eg absorption spectra BFL E.
. . *Conditions, influencing factors*
BFE 94J . . . Pressure
94P . . . Temperature
94Y . . . Electrical & magnetic fields
G . . Continuous spectra
H . . Line spectra
J . . Band spectra
L . . Raman spectra
M . . Mass spectra
. Special attributes in frequency
N . . Pitch
. . . * See Acoustics BRG HFE N
P . . Tone
Q . . Beats
S Amplitude
. Processes
SCN . . Damping

[Forms of motion]
[Periodic motion BDS]
[Waves BF]
[Properties & processes]
[Amplitude BFES]
. [Processes]
. . [Damping BFESCN]

BFE SL . Level (amplitude)
T . Small amplitude
U . Medium amplitude
V . Large amplitude
X Phase
Y . Phase conjugation
BFF . Coherence
BFF N . . Coherence time
R . . Coherence length
BFG Emission
BFG E . Emission spectra
EG . . Continuous emission spectra
J . Emissivity
K . Exitance, emittance
. . * For stimulated emission of radiation, see
. . Techniques B6K QM; for excitation, see
. . BFI N; for lasers, see B6K S.
P . . Radiant exitance, radiant emittance
. . . * Radiant flux leaving a surface per unit
. . . area.
. . . * See also Luminous exitance BRL FGO
Q . Incandescence
. . * See BRL FHQ (optics)
R . Prompt emission, delayed emission
. . * See BOF GR (nuclei)
T . Secondary emission, re-emission
U . Thermionic emission
. . * See BQU FGU (ions)
BFH . Luminescence
. . * See electromagnetic radiation BKF H; light
. . luminescence BRL FH
BFH V Wavetrain
W Wavefront, primary wavefront
. * See also Rays BGC
X Secondary waves, wavelets
. * For secondary emissions, see BFG T.
. Theory
X8I . . Huygen's principle
BFI Transparence
. * For optical transparence, see BRL FI
BFI K Translucence
L Opacity
. * Reciprocal of transmittance.
N Excitation, stimulation
. * See also Collision (particles) BMF S
Q . Deexcitation
S . Self excitation
T . Excited state
U . . Metastable state
. . . * For example, in radiowaves.

Waves

[Forms of motion]
[Periodic motion BDS]
[Waves BF]
[Properties & processes]
[Excitation BFIN]
. [Excited state BFIT]
. . [Metastable state BFIU]

BFI V . Cascade reaction
. . * See BMF IV (particles)
BFJ Decay
. * For damping (waves) see BFC N
BFJ 9CN . Half life
BFL Absorption
BFL E . Absorption spectra
E2Q K . . Absorption coefficient
. . . * For extinction coefficient, see C Chemistry.
M . Absorptance, absorptivity
MP . . Spectral absorptance
MT . . Internal absorptance
S Saturation
SFL . Saturable absorption
X Refraction & reflection (together)
BFM . Refraction
. . * When wave crosses boundary betweeen two
. . media in which its phase velocity differs.
. . Modulus
BFM 9BI . . . Refractive index
H . . Refractivity
I . . Anomalous refraction
J . . Double refraction (general,
. . birefringence (general)
. . . * See also Polarization BFP
K . . . Ordinary rays
L . . . Extraordinary rays
. . *By medium*
R . . . Refraction in real medium
S . . . Refraction in ideal medium,
. . . standard refraction
T . . Subrefraction
U . . Super-refraction
BFN . Reflection
BFN H . . Reflectance, reflectivity
I . . . Specular reflectance
J . . . Diffuse reflectivity, non-specular
. . . reflection
K . . . Total reflectivity
L . . Total internal reflection
LH . . . Total internal reflectivity
M . . Surface reflection
. . . * For Fresnel reflection, see Optics
. . . BRL FNM R.
N . . Reverberation
. . . * See Acoustics BRG HFN R
Q Imaging
. * Formation of a figure of an object by reflected
. or refracted rays.
. * See also Imaging (techniques) B7I

[Dynamics BCX]
[Kinematics BDA]
[Forms of motion]
[Periodic motion BDS]
[Waves BF]
[Properties & processes]
[Imaging BFNQ]

BFO Resonance
. * Maximum response to a mechanical
. system undergoing forced vibration
. by periodic forces at or near to the
. natural frequency of the system.
. * See also Magnetic resonance BJF O
. Frequency
BFO D . . Resonant frequency
H . Double resonance
L . Combination resonance
S . Dissonance
. . * See Acoustics BRG HFO S
BFP Polarization (waves)
. * Special to transverse waves.
. * For Birefringence, see Double
. refractions BFM J; for Electric
. polarization, see Dielectrics BVI SFP.
BFP R . Relaxation time
S . Plane polarization
T . Circular polarization,
. circulatory polarization
U . Elliptical polarization
Y Interactions (radiation)
BFQ . Diffraction
BFQ T . . Fresnel diffraction
V . . Fraunhofer diffraction
BFR . Interference
. . * See also Standing waves BFY G
BFR H . . Beat frequency
J . . Interference patterns
L . . . Interference fringes
M . . . Moire effect
N . . . Inclusion pattern interference
U . Coupling
. . * Interaction between different
. . properties of a system or between
. . two or more systems.
X . Inclusive interactions
BFS . Collision
. . * Interaction in which momentum is
. . conserved.
. . * For capture, see particles
. . BMF UR.
BFS 9DR . . Cross section (collisions)
. . . * More detail is given at
. . . BMF S9D R (collision between
. . . particles).
L . . Elastic collision
M . . Inelastic collision
T . . Action at a distance

Waves

[Forms of motion]
[Periodic motion BDS]
[Waves BF]
[Properties & processes]
. [Interactions radiation BFPY]
. . [Collision BFS]
. . . [Action at a distance BFST]

BFT . . Scattering
. . . * Irregular reflection or dispersal of waves or
. . . particles.
. . . * See also Collision (particles) BMF S
. . . Propagation loss
BFT CG . . . . Scattering losses(
H . . . Shadows
. . . . * Interference of incident & scattered waves.
. . . *Types of scattering*
. . . . * These are usually special to a particular
. . . . radiation or particle. This class is amplified for
. . . . particles at BMF T and for light at BRL FT
. . . . and details from these may be added here if
. . . . necessary:
. . . . * Add to BFT letters J/Y following BRL FT
. . . . and BMF T; eg
JS . . . . Coherent scattering
BFU F . . . Diffusion (waves)
. . . . * For diffusion in general, see Transport
. . . . processes BAV.
G . . . . Self diffusion
H . . Dispersion
R . . Capture
. . . * See Particles BMF UR
BFV . . Annihilation
. . . * See BMF V (particles)
Types of waves
. * For waves defined by energy form, see latter (eg
. electromagnetic waves BKL/BL).
BFV V . *By property*
. . * Add to BFV V letters D/V following BF; eg
VEV . . Large amplitude waves
VP . . Polarized waves
. *By directional & transience factors*
BFW C . . Isotropic waves
D . . Anisotropic waves
L . . Waveforms
LT . . . Correction
M . . . Sine waves, sinusoidal waves
N . . . Square waveform
P . . . Sawtooth waveform
Q . . . Rectangular waveform
R . . . Spherical waveform
S . . . Cylindrical waveform
V . . Wavegroups
. . . * Intermediate between pulses & pure sine waves.
BFX . . . Pulses (wave motion)
BFX 9B . . . . Dimensions
9DJ . . . . . Pulse width
FD . . . . Frequency

[Periodic motion BDS]
[Waves BF]
[Types of waves]
[By directional & transience factors]
[Wavegroups BFWV]
. [Pulses wave motion BFX]
. . [Frequency BFXFD]

BFX FDH . . . Repetition frequency
Q . . Giant pulses
BFY B Time dependent waves
C Random waves
D Torsional waves
E Plane waves
F Oscillating waves
G Standing waves, stationary waves
. * See also Interference BFR
GFQ . Diffraction
H Continuous waves
J Linear waves
L Non-linear waves
N . Solitons, solitary waves
. . * See also magnetic monopoles BNI
P Rotational waves
R Transverse waves (general)
S . Shear waves
U Longitudinal waves (general)
V . Axial vibration
BGA Shock waves, compression waves
. * Usually refer to bodies moving at high speed
. in a compressible fluid. See BSG A for a
. detailed schedule.
BGA R Blast waves
BGB Beams (radiation/waves)
. * Radiation travelling nearly unidirectionally.
. * For particle beam handling, see Imaging
. BM7 I.
. *Operations*
BGB 7IE . . Focusing (beams)
7IF G . . Modulating (beams)
7IF J . . Pulsing (beams)
7IF Q . . Collimation
. *Types*
Q . . Reference beams
R . . Crossed beams
S . . Low energy beams
T . . Refractory beams
TV . . . Doubly refractory beams
BGC Rays, rectlinear propagation (wavefront)
. * Geometrical path normal to wavefront.

[Energy interactions & forms BAF]
[Mechanics BB]
. . . . [Periodic motion BDS]
. . . . . [Waves BF]
. . . . . . [Types of waves]
. . . . [By directional & transience factors]
. . . . [Rays BGC]

BGD . . . . . . *Other types of waves*
. . . . * Should any other concepts from the schedule in bulk matter prove applicable here, proceed as follows:
. . . . * Add to BGD letters GC/M following BS (Fluids); eg ripple waves BGD GNR.
BGH . . . . . . Acoustics
. . . . * See bulk matter BRG H
BGP . . . . . . Thermal properties
. . . . * See Bulk matter BRG P
BGR Gravitation, gravity
. * For Weight, see Statics BCK.
. * See also General relativity B8J
BGR 2M . Mathematics
2P2 L . . Lagrangians (gravitation)
. Constants
76C . . Gravitational constants
. Particular theories
8FG . . Unified field theory
8FL . . Supersymmetry theory (gravity)
. . . * For gravitons, see Elementary particles BMP GN
8H . . Relativity theory
8M . . Quantum theory of gravitation
8MF . . Quantum field theory of gravitation
8ST . . . Supergravity
8TB . . Falling body theory
8TD . . Continuous media gravity
8TE . . . Mixed gravitational systems
8TL . . Alternative theories of gravity
9GV . Conservation laws
9KX R . Many-body problem
BB . Energy
BBP . . Gravitational potential
. Force
BH . . Gravitational field
. . . * For ballistics, see BGR G.
CS . Motion
DA . . Kinematics
DD . . . Gravitational acceleration
. . Waves
F . . Gravitational waves, gravitational radiation
. . . * For gravitons, see Elementary particles BMP GO.
G . . Ballistics (gravitational field)
. . . * Study of the dynamics of the path taken by an object moving under the influence of a gravitational field.

[Physics B]
[Energy interactions & forms BAF]
. [Gravitation BGR]
. . [Motion BGRCS]
. . . [Ballistics gravitational field BGRG]

. . . . * For ballistics as a form of motion, see BDL V.
. . . . * See also Impact BBJ U
BGR GJ . . . . Trajectories (gravitation)
GL . . . Free fall (gravitation)
. . *Properties, effects*
H . . . Centre of gravity
H9D F . . . . Centre of gravity of lines
H9D P . . . . Centre of gravity of planes
H9D S . . . . Centre of gravity of volumes
L . . Absolute gravity
Special energy forms
BGX . Relations between energy forms

BGY . Electricity & magnetism
. . * Field, wave, particle and dynamic aspects, not necessarily associated with a particular state of matter.
. . * Many major concepts of electricity and magnetism are dependent on bulk matter and most of the literature implies it (and in particular, its solid state). So the main schedule is given under Solid state at BVH. The detail there may be used as required here and under each of the states of matter at BR/BW.
. . * Add to B letters GY/J following BV when applicable.
. . * For electromagnetic waves and radiation, see BK.
BH . . Electromagnetism
. . . * Study of electric and magnetic fields and their interaction with electric charges and currents.
. . . * For electromagnetism as a form of magnetism, see BVJ PX.
BHB H . . . Electromagnetic field
. . . . * For electric fields, see BHI BH; for magnetic fields, see BJB H.
. . . . * See also Electrostatics BHN
. . . . *Properties*
HJ . . . . . Gradient
HK . . . . . Field strength
HL . . . . . Field interaction
HM . . . . . Excitation
HN . . . . . Deexcitation
HO . . . . . Field effects
BHI . . Electricity, electrical properties
BHI 92H . . . Variables
. . . . * Use BHJ C.
BH . . . Electric field
. . . . * Usually implies electrostatic field; see BHN BH.

Electricity & magnetism

[Physics B]
[Special energy forms]
[Electricity & magnetism BGY]
[Electricity BHI]
[Electric field BHIBH]

BHJ C Electrical quantities, electrical variables
. * For electric intensity, see electric field strength BVH JN.
. *Input-output relations*
G . . Gain
L . . Loss
BHK . Charge, electric charge
BHK L . . Electrification
. . . * See also Electrostatics BHN
NT . . Attraction
NV . . Repulsion
O . . Electric moments
P . . Polarity
Q . . Electric dipoles
R . . Positive charge
S . . Negative charge
V . . Hypercharge
BHL . Voltage, potential difference, electromotive force, EMF
BHL X . Capacitance
BHM C . Electrical power
. . * See bulk matter BRH MC. The detailed schedule is under solid state (BVH MC).
BHN Electrostatics, static electricity
. * Phenomena associated with electric charge at rest, as compared with current electricity.
. * For electrodynamics, see BHO.
BHN BG . Electrostatic forces
BH . . Electrostatic field
. . . * See note at BHI BH.
. Charge
K . . Electrostatic charge
KL . Electrification
. . * For Frictional electrification, see BVH NKN
. . Electrostatic induction
KM . . . Inductive electrification
. *Special field properties*
N . . Edge effect
BHO Electrodynamics
BHO P . Electromagnetic induction
. . * For alternating currents, see BVH Y.
R . . Inductance
S . . . Self inductance
U . . . Mutual inductance
BHP . Current electricity
. . * See bulk material, BVH P
BHS . . Circuits
BHS X . . . Admittance
BHU . . . . Conduction (electrodynamics)
BHV . . . Impedance
BHW B . . . . Resistance (electrodynamics)
. . *Types of currents*

[Physics B]
[Special energy forms]
[Electricity & magnetism BGY]
[Electricity BHI]
. [Electrodynamics BHO]
. . [Current electricity BHP]
. . . [Types of currents]

. . . . * See Bulk matter BVH XB.
BIU . *Interactions of electricity with other energy forms*
. . * For interactions in which electricity is the agent & not the recipient of the action, see the latter; eg electrooptics BLH (in which it is the optical properties which are affected).
. . * Add to BI letters U/V following BVI; eg
BIV . . Photoelectric effect, photoelectricity
BJ Magnetism
. * See note at BGY re magnetism and bulk matter.
. * Add to BJ letters A/Y following BVJ where applicable; eg
BJB H . Magnetic field
. *Processes & properties*
IM . . Magnetic moment, dipole moment
BJC P . . Relaxation
BJF O . . Resonance
BJK . . Magnetic flux
BJK S . . . Susceptibility
BJL . . Magnetization
. *Field components*
BJN . . Magnetic monopoles
BJN Q . . Dipoles
. *Forms of magnetism*
BJQ . . Diamagnetism
BJR . . Paramagnetism
BJU . *Interactions with other energy forms*
BJU B . . Magnetomechanical effects
C . . . Magnetostriction
BK Electromagnetic radiation (general), electromagnetic waves (general)
. * For waves in general, see BF; for audio frequency waves, see Bulk matter BRG H.
. * Many properties & processes applicable to all or most forms of e-m radiation are more usually considered under Optics (see BL). But all the properties, etc. given under waves in general at BF are available here.
. Investigative techniques & agents
BK7 3 . . Production techniques, generation
3K . . . Sources
3KC . . . . Positive sources
3KD . . . . Negative sources
3KE . . . . Ray sources
3KG . . . . . Positive ray sources
3KH . . . . . Negative ray sources
3KM . . . . Particle sources
3KN . . . . . Positive particle sources
3KO . . . . . Negative particle sources

[Electricity & magnetism BGY]
[Electromagnetic radiation general BK]
[Investigative techniques & agents]
. [Production techniques BK73]
. . [Sources BK73K]
. . . [Particle sources BK73KM]
. . . . [Negative particle sources BK73KO]

BK7 3KP . . . . Electrons (electrical sources)
3KQ . . . . Ions (electrical sources)
3KR . . . . . Positive ion sources
3KS . . . . . Negative ion sources
. Measurement
6 . . Radiometry
64 . . . Radiometers
Energy
BKB B . Radiant energy
BKC W . . Radiant flux, radiant power
WL . . . Radiant flux density, irradiance
BKD C Velocity
. * For Speed of light see BLD C.
BKF C Transmission
DL Low frequency
. * For low frequency RF waves, see BKN L.
G Emission
. * For stimulated emission, see Techniques B6K QM.
H . Luminescence
. . * Usually implies optics, BLF H.
Wavefront
HW . Primary radiation
HX . Secondary radiation
P Polarization
T Scattering
TN . Thomson scattering
. . * See also Thermoelectric effect BVI UGP
BKG *Electromagnetic processes & properties*
. * Add to BKG letters H/J following B so far as
. applicable.
Types of radiation by property
. * Add to BKH letters A/Y following BF;
. * Add to BKI letters A/F following BG; eg
BKH F . Coherent radiation
. . * See also maser & laser techniques B6K QM
P . Polarized radiation
PT . . Circulatory polarized radiation
X . Pulse radiation
By product
BKJ . Ionizing radiation (general)
. . * Electromagnetic or particulate radiation which
. . turns a neutral target (particle or bulk matter) into
. . a charged one.
. . * For particular forms, see the wave or particle
. . causing it.
. . * For the process in a particular medium, see medium.
. . * For the subject of the charged target itself, see

[Physics B]
[Special energy forms]
[Electricity & magnetism BGY]
[Electromagnetic radiation general BK]
[By product]
. [Ionizing radiation general BKJ]

. . Ion physics BQU.
. . * See also Class E/H Biology (including
. . human biology and medicine) for ionizing
. . radiation as agent in pathology, therapy, etc;
. . for deionization, see BKJ S.
. . Energy
BKJ BBP . . . Ionization potential
. . Propagation
FC . . . Ionization, ionized state
. . . . * See also states of matter; eg Gases -
. . . . Ionization BTM KJF C
. . . . Collision
FS9 DR . . . . . Ionization cross section
FSI S . . . . . . Ion surface impact
P . . Irradiation
. . . * Exposure to ionizing radiation.
PQ . . . Microirradiation
S . . Deionization
SR . . . Ion recombination
. . *Types of ionization*
T . . . Photoionization
U . . . Cosmic ray ionization
V . . . Low level ionizing radiation
BKK B . Non-ionizing radiation (general)
Types of radiation by energy state
D . Transition radiation
By frequency & wavelength
BKL . Long waves (general)
BKM . Radiowaves, radiofrequency waves,
. Hertzian waves
BKM FC . . Propagation
FCS . . . Atmospheric propagation (RF)
FCT . . . Scatter propagation, horizon
. . . propagation
. . . . * See also Scattering BFT
FCU . . . . Ionospheric propagation (RF), sky
. . . . waves
FCU J . . . . Magnetoionic propagation (RF)
FCV . . . . Tropospheric propagation
FCW . . . Groundwave propagation (RF),
. . . terrestrial propagation (RF)
FCX . . . Multipath propagation (RF), mixed
. . . path propagation (RF)
GB . . . Radio beams
. . . *By origin*
R . . . . Extraterrestrial radiowaves
S . . . . . Solar radiation (RF)
T . . . . . Cosmic radiation (RF), galactic
. . . . . radiowaves
. . *RF waves by specific frequency*
BKN J . . . Audiofrequency radiowaves

[Special energy forms]
[Electricity & magnetism BGY]
[Electromagnetic radiation general BK]
[By frequency & wavelength]
[Radiowaves BKM]
. [RF waves by specific frequency]
. . [Audiofrequency radiowaves BKNJ]

. . . * c. 30 Hz/20kHz.
BKN K . . Very low frequency waves, myriametric waves, VLF waves
. . . * c. 30-3 kHz.
L . . Low frequency waves, kilometric waves, LF waves
M . . Heterodyne frequency
N . . Medium frequency, MF waves, hectometric waves
BKO . . High frequency, HF waves, short waves, decametric waves
BKP . . Very high frequency, VHF waves, metric waves, millimetre waves
BKQ Microwave & optical physics (together)
. * For quantum optics, see BL8 M; for stimulated emission devices (masers and lasers), see physical techniques B6K QM.
BKU . Microwaves, mm waves, submillimetre waves
. . * From very short wave RF to infrared.
BKU P . . P-band waves
S . . Centimetre wave frequency
U . . . Ultrahigh frequency waves, UHF waves
V . . . . L-band waves
X . . . S-band waves
BKV B . . . Superhigh frequency waves, SHF waves
M . . . . C-band waves
N . . . . X-band waves
P . . . J-band waves
Q . . K-band waves
BKW E . . Extremely high frequency waves, EHF, Millimetre wave frequency waves
G . . . Q-band waves
J . . . V-band waves
L . . . O-band waves
N . . . W-band waves
Q . Maximum observable frequency waves
S . Maximum usable frequency waves, MUF waves

[Physics B]
[Special energy forms]
[Electricity & magnetism BGY]
[Electromagnetic radiation general BK]
[By frequency & wavelength]
[Microwave & optical physics together BKQ]
. [Maximum usable frequency waves BKWS]

BL Optics (general)
. * Originally confined to the study of visible light, the term 'optics' is usually used to cover infrared, visible light and unltraviolet. But it is sometimes used to embrace most of the electromagnetic spectrum in one way or another (eg including the behaviour of electrons and neutrons in an optical context). Here, it is restricted to the frequencies from infrared to ultraviolet.
. * Many of the concepts are dependent on the optical properties of bulk matter. Also, most of the literature refers to visible light in that context (BLV). So the detailed schedule is given at BRL (Bulk matter optics) and details from that may be added here as necessary.
. * An alternative (not recommended) is to confound visible light with optics in general and use BRL 2/BRL Q for it (BL2/BLQ when applied here).
. * Add to BL numbers & letters 2/Q following BRL if applicable.
BL2 M . Mathematics
M94 . . Fourier transform optics
BL3 6 . Practical & experimental optics
. . * A detailed schedule is given at BRL 36.
BL4 . Optical instruments
. Theories
BL8 D . . Emission theory of light (Newton)
DS . . Wave theory of light (Huyghen, etc.)
FDM . . Electromagnetic theory of light (Maxwell)
M . . Quantum optics (general)
BL9 M . . Non-linear optics (general)
BLA F . Energy interactions & forms
BLB B . . Energy
BLD A . . Kinematics
. . . Velocity
C . . . . Speed of light
. . *Special radiation properties & processes*
. . . * For optical properties of materials (eg optical activity, optical rotation) see Bulk material BRL.

[Special energy forms]
[Electricity & magnetism BGY]
[Electromagnetic radiation general BK]
[By frequency & wavelength]
[Optics general BL]
[Energy interactions & forms BLAF]
[Special radiation properties & processes]

Wave motion
BLF . Physical optics (general), electromagnetic optics (general), light waves
. . * Light as electromagnetic waves; for geometrical optics, see Rays BLG C.
BLF C . Propagation, transmission
. . * For optical fibres, see Optical transmission techniques B6L FC.
CM . Attenuation
D . Frequency
E . Spectra
EX . Phase
EY . . Optical phase conjugation, wave front reversal, reversed reflection
F . Coherence
. . * For coherent light, see BLP F.
G . Emission
H . Luminescence
. . * See Bulk matter BRL FH
L . Absorption
LX . Refraction & reflection together
M . . Refraction
MJ . . . Double refraction, birefringence
. . . . * See also Electrooptics BLH
ML . . . . Extraordinary rays
N . . Reflection
P . Polarization
PR . . Optical rotation
Q . Diffraction
R . Interference
T . Scattering
UF . Diffusion
UP . Dispersion
. *Types of light motion*
WR . . Spherical waves
WS . . Cylindrical waves
YN . . Optical solitons, solitary waves (optics)
. . Rays

[Electricity & magnetism BGY]
[Optics general BL]
[Energy interactions & forms BLAF]
. [Special radiation properties & processes]
. . [Wave motion]
. . . [Types of light motion]
. . . . [Rays]

BLG C . . . . . Geometrical optics (general), rectilinear propagation (optics)
. . . . . . * Assumes rectilinear propagation of light, as rays, without reference to waves or the physical nature of light. The behaviour of light in optical instruments.
. . . . . . * See also Optical instruments BRL 4
. *Interactions with other energy forms*
. . * For acoustooptics, see Bulk matter BRL GH.
BLH . . Electrooptics, electric double refraction, optoelectronics
. . . * See also Dielectrics BVI S
BLJ . . Magneto-optical effects
. *Special optical properties & processes*
BLL . . Luminosity
. . . * See Bulk matter BRL L
BLM . . Colour
. . . * See Bulk matter BRL M
. *Other special optical properties*
BLO B . . Optical bistability
C . . Optical multistability
. *Types of light by property*
. . * Add to BLP letters C/Y following BF.
. . * Add to BLP Q letters A/D following BG; eg
BLP F . . Coherent light, coherent optics
. . . * For laser techniques, see B6K S.
M . . Refracted light
N . . Reflected light
P . . Polarized light
PP . . . Partially polarized light
PT . . . Circularly polarized light
PU . . . Elliptically polarized light
WC . . Isotropic light
WD . . Anisotropic light
X . . Light pulses
YH . . Continuous light waves
BLQ B . . Beams (light)
. *By source*
N . . Natural light
P . . Artificial light
. *By frequency & wavelength*
BLU . . Infrared radiation
BLU FE . . . Spectra
N . . . Near infrared radiation
R . . . Far infrared radiation
BLV . . Light, visible light
. . . * See notes at BL above.
. . . * For photons, see BNG O.
BLW Ultraviolet radiation
BLW N . Long wave ultraviolet radiation

[Physics B]
[Special energy forms]
. [Electricity & magnetism BGY]
. . . [By frequency & wavelength]
. . . . [Optics general BL]
. . . . . [Ultraviolet radiation BLW]
. . . . . . [Long wave ultraviolet radiation BLWN]

BLW V . . . . . . Vacuum ultraviolet radiation, far ultraviolet radiation
Y . . . . Radiology
. . . . . * Study of penetrating ionizing (or non-ionizing) radiation.
BLX . . . . . X-rays, Roentgen rays
. . . . . . * Penetrating electro-magnetic radiation, usually generated by accelerating electrons to bombard a solid body, or by inner shell transition of atoms.
BLX FC . . . . . . Production
FE . . . . . . Spectra
FL . . . . . . Absorption
FN . . . . . . Reflection
FQ . . . . . . Diffraction
FT . . . . . . Scattering
N . . . . . . Grenz rays, Infra-Roentgen rays
P . . . . . . Continuous X-rays, Bremsstrahlung
Q . . . . Cosmic ray X-rays
. . . . * For cosmic rays in physics (general), see BND C.
T . . . . . Gamma radiation
. . . . . . * High energy photons, especially as emitted by a nucleus in a transition between two energy levels.
U . . . . . . Non-solar gamma radiation
BLY Matter
. * For matter and energy treated together, see BAE.
BM . Particles, particle physics, high energy physics-
. . * Terminology varies. Sometimes, particle physics is used to mean elementary particles only and nuclear physics to mean the physics of the single nucleus per se (its structure and the reactions involving changes in it). But sometimes, both terms (especially nuclear physics) are used to cover both the nucleus & elementary particles. Use this class only for works reflecting this wider meaning.
BM2 M . . Mathematical models
SA . . . Groups
BM3 2 . . Research operations general
. . Practical & experimental particle physics
6 . . . High energy physics (experimental physics)
. . . . * Sometimes used more broadly, as synonymous with particle physics. In such cases, use BM.
7 . . . Unwanted effects & safety precautions
B . . . Equipment & materials in general
. . . . *Operation on*
BD . . . . . Design

[Physics B]
[Matter BLY]
[Particles BM]
[Practical & experimental particle physics]
[Equipment & materials in general BM3B]
. [Operation on]
. . [Design BM3BD]

BM3 C . . Handling techniques
R . Materials
U . Equipment, plant
BM4 . Instrumentation
BM4 5 . . Instrument components
7 . . . Computers
8 . . . Control systems
85W . . . . Filters
85Y E . . . . Energizing units
. . . *By energy system*
AC . . . . Electrical & electronic
AP . . . . . Circuits
AQ . . . . . . Analogue circuits
AR . . . . . . Pulse circuits
ART . . . . Pulse height discriminators, kicksorters
AS . . . . . . Counters circuits
AST . . . . Coincidence counters circuits
AT . . . . . . Ratometers
FL . . . . Electrooptics
FN . . . . . Photomultiplier tubes
. . . *Components by internal function*
K . . . . Switching devices
L . . . . . Transducers
S . . . . Input devices
SS . . . . . Sensors
T . . . . Output devices
TR . . . . . Recorders
BM5 . . *Types of instruments*
Operations in investigation, techniques
. *Serving all other operations*
BM6 3 . . Data processing & recording
9 . Physical methods in investigation
B . . Mechanical techniques
H . . Electromagnetic techniques
IB . . . Electronic techniques
K . . . Radiation techniques
KFT . . . . Scatter techniques
. *By action on phenomena investigated*
BM7 3 . . Particle production
. . . * For acceleration, see BM7 T; for pair production, see BMF UL.
3K . . . Sources
3L . . . . Velocity selectors
3P . . . Separation
4G . . Detecting & indicating together
. . . * See also Visualization & imaging BM8 GY
. . . Detection of particles
4J . . . . Radiation detection of particles
. . . . . * For track visualization, see BM8 H.

[Particles BM]
[Practical & experimental particle physics]
[Operations in investigation]
[By action on phenomena investigated]
[Detecting & indicating together BM74G]
. [Detection of particles]
. . [Radiation detection of particles BM74J]

BM7 4J7 C . . . Monitoring
. . . . * For dosimetry, see BM7 6K7 8S.
4JL . . . Angular correction techniques
4JN . . . Coincidence techniques
5 Counting & detection (particles), counting (particles)
. Instruments
54 . Counters & detectors
. . * Synthesis by AY5 is modified here:
. . * Add to BM7 5A numbers & letters 5/U following AY4 if applicable.
5B . . Gas ionization counters, ionization counters
5C . . . Ionization chamber counters
5CE . . . . Integrating ionization chamber counters
. . . . . * With long response time.
5CF . . . . . Lauritson electroscope
5CH . . . . Non-integrating ionization chamber counters
5D . . . Proportional counters
5E . . . . Geiger-Muller counters
5EG . . . . . Self-quenching Geiger counters
5EJ . . . . . Non-self-quenching Geiger counters
5F . . . . Gas-flow counters
5G . . . Spark counters
5H . . Solid state ionization counters, crystal counters
5J . . . Semiconductor counters (particles), photodiode counters
5K . . . . Intrinsic semiconductor counters
5KN . . . . . NaCl semiconductor counters
5KP . . . . . CdS semiconductor counters
5KR . . . . . Diamond semiconductor counters
5L . . . . Junction semiconductor counters
5LN . . . . . Diffused semiconductor counters
5LS . . . . . Surface barrier semiconductor counters
5M . . . . . Ion drifted semiconductor counters
5MP . . . . . . Lithium drifted semiconductor counters
5MS . . . . . . Silicon semiconductor counters
5MT . . . . . . Germanium semiconductor counters
5N . . Scintillation counters, scintillators
5N5 . . . Hodoscopes
5NP . . . Inorganic activated scintillators
5NR . . . Organic activated scintillators

[Particles BM]
[Operations in investigation]
[By action on phenomena investigated]
[Counting & detection particles BM75]
. [Counters & detectors BM754]
. . [Scintillation counters BM75N]
. . . [Organic activated scintillators BM75NR]

BM7 5NT . . . Liquid scintillation counters
5P . . Four-pi counters
5Q . . Cerenkov counters, Cherenkov counters
5R . . Electron multiplier counters
5S . . Neutron counters
. . . * For scintillation counters, see BM7 5N.
5ST . . . Gas counter with nuclides
5SU . . . . BF3 filled gas counter
5SV . . . . Helium filled gas counters
5SW . . . . Fission chambers
5T . . . Fast neutron detectors
5V . Other detector & counters
. . * Not classifiable under above classes. Arrange A/Z; eg
5VC . . Coincidence counters
5VF . . Foil activation (particle detection)
6 Measurement
6K7 8S . Dosimetry
F Simulation & modelling
Visualization & imaging together
H . Track visualization (particles)
. . Instruments
H5V . . . Hodoscopes
. . *Techniques defined by equipment*
HJ . . . Cloud chambers, Wilson cloud chamber
HK . . . . Diffusion cloud chambers
HL . . . . Expansion chambers (cloud chambers)
HM . . . . . Bubble chambers
HN . . . . . Luminescence chambers, scintillation chambers
HP . . . . . Spark chambers
HPL . . . . . . Filmless spark chambers
HPM . . . . TV camera spark chambers
HPN . . . . Sonic spark chambers
HPP . . . . Wire spark chambers
HQ . . . Emulsion techniques
HU . . . . Nuclear emulsions (particle detection), particle sensitive emulsions, photographic emulsions (particle detection
HV . Visualization techniques
I . Imaging
. . * Add to BM7 I letters C/L following AY7 I (if applicable) for imaging techniques other than by particle beams (for which see BM7 IM).
IM . . Particle beam techniques, particle optics (imaging), beam handling

[Particles BM]
[Practical & experimental particle physics]
[Operations in investigation]
[By action on phenomena investigated]
[Visualization & imaging together]
[Imaging BM7I]
[Particle beam techniques BM7IM]

BM7 IM4	Instruments
IM4 FV	. Lenses (particle optics)
IM4 FVU	. . Quadrupoles
IM4 FVV	. . Electrostatic lenses (particle optics)
IM4 FVW	. . Magnetic lenses (particle optics)
IM4 FX	. Electrodes (particle optics), probes (particle optics)
IM4 FYE	. Beam expanders
IM4 FYG	. Beam resonators
IMC B	*Target materials*
	Operations
IMC N	. Optical transfer function
IMD	. Beam handling
	. . * Use the general classmark BM7 IM.
IMD F	. Beam defection
IMD G	. . Injection (particle optics)
IMD H	. . . Guns (particle optics)
IMD K	. Extracting (particle optics)
IMD L	. . Repetitive cycling
IMD P	. Positioning (particle optics)
IMD Q	. . Alignment (particle optics)
IME	. Focusing (particle optics)
IME G	. . Beam trapping
IME H	. . . Self-trapping
IME J	. . Electrostatic focusing
IME M	. . Magnetic focusing
IME N	. . Bunching
IME P	. . Phase focusing
IME R	. . Radial focusing
IME S	. . Axial focusing
IME T	. . Strong focusing, alternating gradient focusing, AG focusing
IME U	. . Momentum focusing
IME V	. . Velocity focusing
IMF	. Resolution
	. . * For magnification, see AY7 6S.
IMF H	. Modulating (particle optics)
IMF J	. Shaping
IMF L	. Splitting
IMF P	. Separation (particle optics)
IMF Q	. Stacking (particle optics)
IMF S	. Scanning (particle optics)
IMF T	. Projection (particle optics)
IMF V	. . Collimation
IMF W	. Beam pulsing & switching

[Particles BM]
[By action on phenomena investigated]
[Visualization & imaging together]
[Imaging BM7I]
. [Particle beam techniques BM7IM]
. . [Operations]
. . . [Beam pulsing & switching BM7IMFW]

	. . *Properties of beams*
BM7 IMG B	. . . Beam profile
IMG E	. . . Beam diameter
IMG G	. . . Beam angle
IMG J	. . . Beam edges
IMG L	. . . Contrast
IMG M	. . . Divergence (particle beams)
IMG P	. . . Flux density (particle beams)
IMG Q	. . . Emittance (particle beams)
IMG V	. . . Velocity (beam handling)
	. . *By type of beam*
	. . . * Add to BM7 I letters N/Q following B; eg
INP	. . . Electron beams, electron optics, optoelectronics, optical electronics
	. . . . * For optoelectronics, see BRL H.
INP X	. . . . Cathode rays (electron optics)
INV	. . . Proton beams
IO	. . . Nuclear magnetic resonance imaging, NMR imaging, magnetic resonance imaging, MRI, proton resonance imaging
	. . . . * Using magnetic resonance of protons.
IQU	. . . Ion beams (techniques), ion optics
T	Acceleration (particle physics), accelerator techniques
	. * For high-energy physics, see Particles - Practical physics BM3 6.
	. Equipment & plant
T3U	. . Particle accelerators, generators (accelerators)
T3U 3R	. . . Materials
T4A X	. . . Waveguides
	. . . . * Use BM7 TK
	. . . *Parts of accelerators*
TF	. . . . Particle sources
TG	. . . . . Primary acceleration
	. . . . . . * Using another, simpler accelerator.
TH	. . . . Power supply (accelerators)
THJ	. . . . . Voltage multipliers, voltage amplifiers
	. . . . . . * For Cockcroft-Walton accelerators, see BM7 UK.
TJ	. . . . Magnets (accelerators)
TJP	. . . . Coils (accelerators)
TJV	. . . . Vacuum chambers (accelerators)
TJX	. . . . Storage rings (accelerators)
TK	. . . . Waveguides (accelerators
TKR	. . . . . Resonant cavities
TKS	. . . . . . Superconducting resonant cavities
TL	. . . . Other accessories

Acceleration (particle physics)

[By action on phenomena investigated]
[Visualization & imaging together]
[Acceleration particle physics BM7T]
[Equipment & plant]
[Particle accelerators BM7T3U]
[Parts of accelerators]
. [Other accessories BM7TL]

. . * Arrange A/Z.
Types of accelerators
. * Any given type may be qualified as follows (where the hyphen represents its classmark):
. * Add to - letters F/L following BM7 T
. *By particle accelerated*
. . * The preferred arrangement when a particular particle is accelerated is to locate under that particle (in BN/BQ); eg Positron - Accelerators - Linear BNP RF7 UN.
. . * An alternative (not recommended) is to cite the kind of accelerator first; in this case use this position & proceed as follows (where the hyphen represents the type of accelerator in BM7 TN/V):
. . * Add to -N letters B/Y following BN; eg Linear accelerators - Positron BM7 UNN PRF.
. . * Add to - letters O/Q following B; eg Van de Graaf - Heavy ion accelerator BM7 UHT QUS.
. . *Specific particles*
. . . * For general works on these as objects of acceleration (if the alternative above is followed)).
. . . * Add to BM7 T letters N/Q following B; eg

BM7 TNP . . . Electron accelerators
TNP RF . . . . Positron accelerators
TNV . . . Proton accelerators
TQU . . . Ion accelerators
TQU S . . . . Heavy ion accelerators
. *By shape*
TT . . Linear accelerators (general), linacs
TU . . Orbital accelerators (general)
. *By field*
TW . . Fixed field accelerators
TX . . Varying field accelerators
. *By focusing gradient*
UC . . Weak accelerators
UE . . Strong accelerators, alternating gradient accelerators
. *By push*
UG . . Single push accelerators, electrostatic generators
. . . * Uses very high voltage to give direct acceleration.

[Visualization & imaging together]
[Acceleration particle physics BM7T]
[Equipment & plant]
[Particle accelerators BM7T3U]
[Types of accelerators]
[By push]
[Single push accelerators BM7UG]

BM7 UH . Van de Graaf generator
. . * See also Heavy ion accelerators- - Van de Graaf BQU S7U H.
UJ . Tandem generator
UK . Cockcroft Walton generator
UM Multipush accelerators, resonant accelerators
UN . Linear (multipush) accelerators
. . * See also particular particles accelerated: eg Heavy ion linear accelerators - Wideroe BQU S7U NW.
UO . Orbital accelerators (multipush), cyclic accelerators
UP . . Cyclotrons, fixed field orbital accelerators
. . . * Usually assumed.
UR . . . Isochronous cyclotrons
US . . . . Fixed field alternating gradient cyclotrons
. . . . . * See also Electron cyclotrons BNP 7US; Ion cyclotrons BQU 7US.
UW . . . Synchrocyclotrons, frequency-modulated cyclotrons
. . . . * See also Ion synchrocyclotrons BQU 7UW
UX . . . Microtrons
. . . . * See also Electron microtrons BNP 7UX
VC . . Synchrotons, varying field orbital accelerators
. . . * See also Electron synchrotons BNP 7VC; Proton synchrotons BNV 7VC
VG . . . Alternating gradient synchrotons
. . . . * For Betatrons, see Electron synchrotons BNP 7VG; for Fixed field alternating gradient ring accelerators, see BNV 7VH.
VJ . . . Bevatrons
VK . . . Cosmotrons
VL . . Ring accelerators
. . . * Induction accelerated.
VM . . Separated orbit accelerators
. . . * Usually for protons.

[Practical & experimental particle physics]
[Operations in investigation]
. . . [Acceleration particle physics BM7T]
. . . . [By push]
. . . . [Multipush accelerators BM7UM]
. . . . [Orbital accelerators multipush BM7UO]
. . . . [Separated orbit accelerators BM7VM]

BM7 VN . . . . Colliding beam accelerators
VNJ X . . . . Storage rings
BM8 B Theoretical physics
F . Field theories
FG . . Unification field theories, unified theories
FH . . . Standard model
FJ . . . Grand unified theories, GUTS
. . . . * Of the four fundamental interactions. May or may not include gravitational force.
FL . . . Supersymmetry grand unified theories
. . . . * Includes gravitationl force.
FR . . Gauge theory
. . . * For Weinberg model, see electroweak interactions, BMP J3.
FS . . . Gauge transformations
FT . . . Gauge invariance
GP . Non-field theories
GS . . String theories
GT . . . Membrane theories
GU . . . Superstring theories
H . Relativity theory
HFG . . Unification theory
M . Quantum theory
. . * For quantum number properties, see BMM D.
MF . . Quantum field theory
. . . *Properties*
MFM . . . . Renormalization
MFN . . . Non-linear quantum theory, non-local quantum theory
MFR . . . Gauge theory
MFW . . . . Asymptotic freedom (gauge theory)
MFX . . . Axiomatic quantum field theory
MGC . . . Schwinger source theory
MGO . . . Nuclear field theory
MJF . . . Relativistic quantum field theory
. . . . * See also quantum electrodynamics BNR
. Other special theories & models
SB . . Diffraction model
SC . . Composite models
SE . . Many body theory
SG . . Spirality theory
SH . . Helicity theory
SJ . . Form factors (elementary particles)
SL . . . Bootstrap theory, bootstrapping
SN . . Duality models, dual models, dual resonance models
SNP . . . Veneziano model

[Matter BLY]
[Particles BM]
[Practical & experimental particle physics]
. [Theoretical physics BM8B]
. . [Other special theories & models]
. . . [Duality models BM8SN]
. . . . [Veneziano model BM8SNP]

BM8 SP . . . Peripheral models, exchange models
SPQ . . . . Multiperipheral models, multi-Regge models
SR . . . Regge poles & trajectories
SS . . . Pomeranchuk poles & trajectories
General processes/properties in particle physics
BM9 C . Time
CI . . Lifetime
. . . * See quantum properties BMM FH
D . Space
. . * See also quantum properties parity BMM E and spin BMM K.
DJ . . Width
DK . . Radius
G . Systems characteristics
GV . . Conservation laws
. . . * See also Symmetry & conservation BMM AI; Equivalence of energy & matter BAE 9GW
J . . Structure
JS . . . Symmetry
. . . . * Use symmetry as a quantized property (BMM B).
Energy interactions & forms
BMA F . Interactions of particles (general)
. . * Exchange of energy between two particles or between a particle and an electromagnetic wave.
. . * For interactions between two particles, see BMP
F2X . . Statistical models
F8B . . Theory
. . . * Theories & models centred on particular types of particles, properties, etc. go with the particle, property, etc.; eg QCD (quantum chromodynamics) goes umder quarks (at BNR 8M).
G . . Thermodynamics
T . . . Transport processes
BMB . . Mechanics

Energy

[Physics B]
[Matter BLY]
[Particles BM]
[Energy interactions & forms]
[Interactions of particles general BMAF]
[Mechanics BMB]

BMB B Energy
. * For energy associated with spin, see
. Spin BMM K.
C . Energy ranges, energies
. (particles)
. . * See also Energy loss BMC F
CD . . Low & intermediate energy
. . ranges
. . . * < 1 GeV.
CE . . . Thermal energies
. . . . * < 5 eV.
CF . . . Resonance energies
. . . . * 5 eV-0.01 MeV.
CG . . . Fast energies
. . . . * 0.01-30 MeV.
CH . . High energies
. . . * 30-100 MeV.
CI . . Very high energies
. . . * > 1 GeV.
CJ . . Superhigh energies
. . . * > 1 TeV.
D . Energy levels (general),
. electron energy states
. . * The detailed schedule for this class
. . is given under atoms, at BPB D.
. . Details from it may be used to
. . qualify this more general class or
. . any particular particle or group of
. . particles to which its concepts
. . might apply.
. . * For energy levels of particular
. . particles, see the particle; eg
. . Molecules - Charge transfer state
. . BQB EW;
. . * For energy bands, see condensed
. . matter BTX BF.
. . * Add to BMB letters D/F following
. . BPB; eg
. . *Properties*
D9D J . . . Energy level width
. . . . * Measure of the uncertainty of
. . . . a specified level.
. . *Processes*
DN . . . Transitions (energy levels)
DQ . . . Isomerism (energy levels)
. . *Types of energy states*
. . . * For types special to a particle,
. . . etc. see the

[Particles BM]
[Energy interactions & forms]
[Interactions of particles general BMAF]
[Mechanics BMB]
[Energy BMBB]
. [Energy levels general BMBD]
. . [Types of energy states]

. . . particle.
BMB E . . . Stationary state, quantum state,
. . . eigenstate, energy eigenstate,
. . . energy state
EG . . . Ground state, lowest energy state
EK . . . Excited state
EM . . . Bound state, discrete energy level
EP . . . Unbound state
EQ . . . Degenerate level
. . . . * Reflecting more than one quantum state.
EQP . . . . Degeneracy
. . . . . * When two or more quantum states
. . . . . have the same energy.
ER . . . Fine state, fine structure
ES . . . Hyperfine state
. . . . * When a magnetic field is applied.
EU . . . Multiplets (energy levels)
. . . . * Sets of related, closely spaced energy
. . . . levels or quantum states.
. . . . * See also multiplets treated as groups of
. . . . particles; eg isospin multiplets (strong
. . . . interactions) BNQ T.
EUK . . . . Spin-orbit multiplets
EV . . . . Doublets state
EW . . . . Triplets state
F . Energy bands [
. . * See Condensed matter BTX BF
. Energy loss of particles
BMC F . . Energy-range relations, particle range
. . (energy loss)
. . . * See also Electron energy loss spectroscopy
. . . B7M NPQ; Channelling effect BMD UL
FH . . . Stopping power
. . . . * Measure of energy loss of substance
. . . . when a charged particle passes through it.
H Statics
J . Mass
JL . . Mass difference
L . Density
X Dynamics
BMD C . Velocity
D . Acceleration
K . *Forms of motion*
MW . . Orbits
N . . Rotation
. . . Momentum
NCV . . . . Angular momentum
NCX . . . . . Complex angular momentum
. . . . . plane
NCX 8SR . . . . . . Regge poles & trajectories

[Particles BM]
[Dynamics BMCX]
[Forms of motion BMDK]
. [Rotation BMDN]
. . [Momentum]
. . . . [Complex angular momentum plane BMDNCX]
. . . . . [Regge poles & trajectories BMDNCX8SR]

BMD NCX 8SS	. . . . . Pomeranchuk poles & trajectories
NS	. . Spin
	. . . * See quantum numbers properties of
	. . . particles BMM K
U	Oscillation
UL	. Channelling effect
ULD K	. . Transients
BME XH	. Forced vibrations
Y	Radiation
BMF	Wave properties
BMF 2M8 L	. Wave function
	. . * For parity, see BMM E; for molecular
	. . orbitals, see BQB DT.
C	. Transmission, propagation
	. . * For particle production, see BMF UJ.
CM	. Attenuation
F	. Coherence
G	. Emission
IN	. . Excitation
	. . . * For excited state, see Energy states
	. . . BMB EH.
	. . . * See also Polaritons BVE RU
IP	. . . Coulomb excitation
IV	. . . Cascade reactions
J	. Decay
	. . * Transformation into a more stable particle.
	. . * For radioactivity, see Nuclear interaction
	. . BOF K.
JFC N	. . Damping
KQF	. . Half life
L	. Absorption
	. . * See also inelastic scattering BMF TM
N	. Reflection
O	. Resonance
	. . * For magnetic resonance, see BVJ FO; for
	. . resonances as types of particles, see BND T.
P	. Polarization
RU	. Coupling
	. . * For K-coupling, see under Capture BMF UQ.
RV	. . Russell-Saunders coupling, L-S
	. . coupling
RVV	. . . Intermediate coupling
RW	. . J-J coupling
RX	. . Inclusive interaction (coupling)
RY	. . Exclusive interaction (coupling)
S	. Collision
	. . * Nearness of approach producing mutual
	. . interaction

[Particles BM]
[Energy interactions & forms]
[Interactions of particles general BMAF]
[Mechanics BMB]
[Dynamics BMCX]
[Wave properties BMF]
[Collision BMFS]

	. short of impact. For actual contact,
	. see capture BMF UM.
	. * When qualifying specific particles,
	. use Scattering BMF T; eg Beta ray -
	. Scattering BNP RDO FT.
	. * For annihilation, see BMF V.
BMF S9D H	. Mean free path
	. . * Usually implies molecules.
S9D R	. Cross section
	. . * Area presented to the incident
	. . particle as a measure of the
	. . probability of a particular collision
	. . process
S9D RQ	. . Integral cross section
S9D RR	. . Total cross section
SJ	. Impact
SL	. Elastic collision
SM	. Inelastic collision
SN	. . Collisions of the first kind,
	. . endoergic collisions
SO	. . Collisions of the second kind,
	. . exoergic collisions
SP	. One-dimensional collision
SQ	. Binary collision
T	Scattering
	. Matrix algebra
T2T B	. . S-matrix theory
T8H	. Relativistic scattering theory
	. *Processes & properties*
	. . * See also Cross-section BMF SH
TCG	. . Scattering losses
TEP	. . Amplitude
TH	. . Shadows
	. *Types of scattering*
	. . * For the scattering of particular
	. . radiations or particles, see latter; eg
	. . Light - Rayleigh scattering
	. . BRL FTQ; Photons -
	. . Scattering - Compton effect
	. . BNG OFT.
TJC	. . Angular distribution
	. . (scattering)
TJF	. . Forward scattering
TJH	. . Back scattering
TJL	. . Elastic scattering
TJM	. . Quasi-elastic scattering
TJN	. . Inelastic scattering,
	. . absorption (scattering)
TJP	. . Multichannel scattering,
	. . multiple scattering

Particles

[Particles BM]
[Interactions of particles general BMAF]
[Mechanics BMB]
. . [Wave properties BMF]
. . . [Scattering BMFT]
. . . . [Types of scattering]
. . . . . [Multichannel scattering BMFTJP]

BMF TJQ . . . . . Many-body scattering
TJS . . . . . Coherent scattering
TJT . . . . . Incoherent scattering
TJV . . . . . Critical scattering
TK . . . . . Resonance scattering
. . . . . . * Incident wave penetrates nucleus.
TKT . . . . . . Potential scattering
. . . . * Incident wave reflected at nuclear
. . . . surface.
TM . . . . *Types of scattering special to a*
. . . . *radiation/particle*
. . . . . * See the radiation or particle; eg Light -
. . . . . Stimulated scattering BRL FTM;
. . . . . Charged particles - Coulomb scattering
. . . . . BNG FTS.
UF . . . Diffusion
UH . . . Dispersion
. . . . * Dependence of wave velocity on frequency; a
. . . . property of the medium in which the wave is
. . . . propagated.
. . . . Symmetry
UH9 JS . . . . . Crossing symmetries
UHL . . . . N/D method (dispersion)
UHN . . . . Form factors (dispersion)
UHP . . . . Multivariable dispersion relations
UL . . . . Pair production
. . . . . * See also annihilation BMF V
UM . . . Recombination
. . . . * See charged particles neutralization
. . . . BNH WFU M

UR . . . Capture
. . . . * Acquisition of an additional particle by
. . . . nucleus, atom, molecule or ion.
US . . . . Electron attachment
. . . . . * Capture by atom molecule or ion.
. . . . . * For radiative capture, see BOF KU.
UT . . . . . K-coupling, K-capture
. . . . . . * Capture of K-shell electron.
V . . . Annihilation
. . . . * Conversion, on collision, of a particle and its
. . . . corresponding anti-particle into radiation.
X . . . Pulses
. . . Beams
BMG B . . . . Particle beams
BMH Electromagnetic properties
. * For QED, see BNG 8S.
BMH I . Electrical

[Physics B]
[Matter BLY]
[Particles BM]
[Energy interactions & forms]
[Interactions of particles general BMAF]
. [Electromagnetic properties BMH]
. . [Electrical BMHI]

. . . * For charge, see quantum properties
. . . BMM M.
BMJ . . Magnetic
BMJ BIM . . . Magnetic moment

Special particle processes/properties
BMM . Quantum number properties
. . * The possible values characterizing a
. . property in a physical system when it has
. . been quantized, i.e. the property has been
. . found to take only certain discrete values.
. . * The preferred arrangement is to collect all
. . quantum numbers together here, enumerating
. . the most prominent ones and providing for
. . the others by synthesis (at BMM D9).
. . * An alternative (not recommended) is to
. . locate any property/process quantized at its
. . normal position in BMA/BML and qualify it
. . by the general class for quantum number at
. . B8N; eg charge quantum number
. . BMH KB8 N.
. . * The order of concepts is the same as that in
. . the general schedule (BMA/BML) but the
. . notation is enumerated.
BMM 9 . . *Quantum numbers other than those*
. . *enumerated below*
. . . * Add to BMM A numbers & letters 9/L
. . . following BM as applicable.
AN . Invariance principle
. . * Symmetry & conservation together.
AS . . Conserved properties
AV . . Not-conserved properties
B . . Symmetry, symmetries
. . . * See also general symmetry B9J S
B2M . . . Mathematics
B2S A . . . . Symmetry groups (mathematics)
. . . . . * For SU2, see BNQ MIT; for SU3, see
. . . . . BNQ TT.
B2S J . . . . . Lie groups
B2S KXQ . . . . . . Poincare groups
. . . Theory
B8B . . . . Invariance theory, symmetry law
B8V L . . . . . Lorenz invariance
B8V P . . . . . Poincare invariance
BT . . . Spontaneous symmetry breaking
. . . . * See also Higgs mechanism BMP JNR S
. . . . ; Mass (quantum properties) BMM H
BX . . . Non-linear symmetries, spectrum-
. . . generating symmetries
CC . . . Discrete symmetries

Particles

[Matter BLY]
[Particles BM]
[Energy interactions & forms]
[Special particle processes/properties]
[Invariance principle BMMAN]
[Symmetry BMMB]
. [Discrete symmetries BMMCC]

BMM CD . . Charge conjugation (symmetry)
CE . . Time reversal (symmetry)
CF . . C invariance
CG . . CP invariance
CH . . CPT invariance
CJ . . T invariance
CK . Dynamical symmetries
CL . Chiral symmetries, chirality
. . * Relationship between spin vector & momentum vector, especially of neutrinos.
CP . Supersymmetry
. . * Hypothesizes a corresponding boson for every fermion and vice versa. See notes at BNJ and BNK.
. Unitary symmetry
. . * See Hadrons BNQ MCS
E Parity, space reflection symmetry, P (parity)
. Conservation
E9G V . . P invariance, parity conservation
EG . G-parity
. . * Quantum number associated with elementary particles having zero baryon number and strangeness. Conserved in strong interactions only.
EK . Charge conjugation parity, C-parity
. . * Quantum number associated with elementary particles having zero charge, baryon number and strangeness.
. . * See also mesons BNS
FH Duration (quantum numbers)
FK . Short lived (quantum numbers)
FL . Long lived (quantum numbers)
FM . Mean life (quantum numbers)
FN . Half life (quantum numbers)
G Size (quantum numbers), dimensions (quantum numbers)
GJ . Width (quantum numbers)
GK . Radius (quantum numbers)
H Mass (quantum numbers)
. * For symmetry breaking, see BMM BT.
I Isospin (general), isotopic spin, isospin isobaric spin, i-spin
. * Usually implies hadrons; see BNQ MI.
J Angular momentum (quantum numbers)
K Spin (quantum numbers)
. * For isospin, see BMQ I
KB8 P . Wave mechanics

[Matter BLY]
[Particles BM]
[Energy interactions & forms]
[Special particle processes/properties]
[Invariance principle BMMAN]
. [Spin quantum numbers BMMK]
. . [Wave mechanics BMMKB8P]

. . . Hamiltonian functions
BMM KB8 P2M 8NH . . . . Spin Hamiltonians
KFT U . . Diffusion
KFT W . . . Spin disorder resistivity
KME . . Parity
KR . . Helicity
KS . . Spin orbit interaction
KW . . Non-zero spin
L . Decay (quantum numbers)
M . Charge (quantum numbers)
. . * For charge conjugation parity, see BMM EK.
MP . . Electrical moment (quantum numbers)
MT . . Hypercharge (quantum numbers)
. . . * Sum of baryon number and strangeness.
NJ . Magnetic moment (quantum numbers)
O . Magnetic resonance (quantum numbers)
. . * Add to BMM O letters P/R following BOM O.
. *Properties defined by particles in system*
. . * Add to BMM P letters B/Y following BN; eg
PM . . Lepton number
PT . . Baryon number
R . *Properties special to particular particles*
. . * This position allows features of particles to be treated as properties (for use as qualifiers) as distinct from using R (in instruction at BNB) to introduce them as specifiersr; eg Resonances - Charm (property) BND TMR U; but Resonances - Charmed (type) BND TRR U.
. . * Add to BMM R letters A/X following BN; eg
RS . . Colour (quantum numbers)
RU . . Charm (quantum numbers)
RW . . Strangeness
Interactions by energy expenditure
. * For inclusive and exclusive interactions, see BMF RX/Y.
. * Add to BMN letters D/J following BMB C with the additions indicated.
BMN J . Superhigh energy interactions

[Physics B]
[Matter BLY]
[Particles BM]
[Energy interactions & forms]
[Special particle processes/properties]
[Interactions by energy expenditure]
. [Superhigh energy interactions BMNJ]

BMN P . Large momentum transfer interactions

V Basic interactions, fundamental interactions

X . Exchange forces, field particle exchange

. . * Forces acting between particles due to exchange of some property (charge, spin, etc).

. . *Agents*

BMO . . . Exchange particles (general), gauge bosons, field particles (general), virtual intermediate bosons, quasi-particles

. . . . * Particles which mediate the interaction between the fundamental particles; carriers of the forces between material particles (see BND X). Their status as particles is somewhat problematical; they are sometimes considered as being constituents of the material particle and their action has been likened to that of pulses. Their general nature is closely dependent on the interactions they mediate and this is the reason for their location here.

. . . . * For particular exchange forces, see the type of interaction; eg Electomagnetic interactions - Photons BNG O.

BMP G . Gravitational interactions

. . * Generally speaking, particles are too small to be subject to gravitational forces. This position is provided for documents which consider these forces in relation to the other fundamental forces.

. . * For gravitation in general, see BGR.

. . Exchange particles

GO . . . Gravitons

J . Electroweak forces, electroweak interactions

J8W . . Weinberg-Salam theory

JHP . . Current

JHP 2RS . . . Current algebras (electroweak forces)

JHP 8B . . . Current theory (electroweak forces)

[Physics B]
[Particles BM]
[Energy interactions & forms]
. [Special particle processes/properties]
. . [Basic interactions BMNV]
. . . [Electroweak forces BMPJ]
. . . . [Current BMPJHP]
. . . . . [Current theory electroweak forces BMPJHP8B]

BMP JHY X . . . . . Neutral weak currents, neutral current interaction

JM . . . . Quantum numbers

JMP M . . . . . Lepton number

. . . . Exchange particles

JO . . . . . W-bosons, intermediate bosons, intermediate vector bosons, IVBs, intermediate mesons

JON . . . . . Z-bosons

JOP . . . . . Higgs boson, Higgs field

JOP Q . . . . . . Higgs mechanisms

. . . . * Generate mass for IVBs.

JOR . . . . . . Goldstone boson

. . . . Electromagnetic forces (elementary particles)

. . . . . * See under charged particles BNG

L . . . . Weak interactions

. . . . . * Only quantum numbers Q (charge) and B (baryon number) are conserved.

LHY X . . . . . Neutral weak currents

. . . . . Exchange particles

LO . . . . . . W-bosons, intermediate bosons

. . . Strong interactions

. . . . * See Hadrons BNQ PN

BMQ . . Interactions with another particle

. . . * This position is used only when qualifying a particular particle.

. . . * When dealing with reactants, the usual rule in BC2 is to cite the product first. However, in particle physics the exact relationships between the reactants are not always easily established and do not provide a basis for a consistent citation order. The rule is therefore to cite first the reactant which files latest in the schedule, followed by the second reactant; eg interaction between leptons (BNM) and hadrons (BNQ) is BNQ QM..

. . . * Add to BMQ letters B/X following BN;

. . . * Add to BMQ Y letters O/Q following B; eg Molecule - Atom interactions BQQ YP.

BMV *Parts of particles*

. * Theoretically, elementary particles cannot exhibit constituent parts. But a number of theories relate to the possible possession of parts.

BMW . Composite particles, composite models

. . * Particles considered to be substructures or constituents of other particles.

. . * See also Exchange particles BMO; Partons BNU RBP; Quarks BNR.

[Physics B]
[Matter BLY]
[Particles BM]
[Parts of particles BMV]
. [Composite particles BMW]

Types of particles
BNB . Elementary particles (types), fundamental particles (types)
. . * Do not qualify this general heading (BNB) by processes and properties, etc.; for these, see BMA/BMQ.
. . * Each type of particle (including its species, if any) may be qualified or specified in detail as follows (where the hyphen represents the particle's classmark):
. . * Add to - numbers & letters 2/9,A/Q following BM (for Processes & properties); eg Electrons - Emission BNP FG.
. . * Add to -QY letters V/W following BM (for Parts)
. . * Add to -R letters B/X following BN (for Types by other types of particle); eg Low energy electrons BNP RLD.
. . * Add to -S letters O/Q following B (continuing Types by other types of particle) if applicable.
. . * (Types special to the particle)
. . * Letters -T/Y are used for types special to the particle; eg BNT T Dibaryons.
BNB B . . *Types by non-quantum property*
. . . * Add to BNB letters B/L following BM if applicable.
V . . *Types by part*
. . . * Add to BNB letters V/X following BM if applicable; at present this is only hypothetical.
BNC . . *Types by special & quantum number property*
. . . * Add to BNC letters B/Y following BMM, except for those indicated below (which are enumerated separately); eg
BNC AS . . . Conserved particles
. . . . * As in strong interactions.
AV . . . Not-conserved particles
. . . . * As in weak interactions.
CP . . . Supersymmetric particles
CQ . . . . Scalar particles
CR . . . . Sparticles
FH . . . *By lifetime, mean life*
. . . . * See BND RH
H . . . *By mass*
. . . . * See BNL Q...
K . . . *By spin*
. . . . * See BNJ...
M . . . *By charge*
. . . . * See BNF X...

[Physics B]
[Matter BLY]
[Particles BM]
[Types of particles]
[Elementary particles types BNB]
[Types by special & quantum number property BNC]
. [By charge BNCM]

By origin. source
. * For works considering the properties, etc of these particles per se (in terms of particle theory). For works regarding the particles as features of their origins, see the relevant context; eg cosmic rays in D Astronomy.
BND A . Particle accelerator particles
. . * Produced by accelerators.
AR . Nuclear reactor particles
B . Atmospheric particles
C . Cosmic rays
. . * See also Very high energy particles BNL G
CFH W . . Primary radiation
CFH X . . Secondary radiation
D . . Showers (cosmic rays), bursts (cosmic rays)
DP . . . Penetrating showers
DQ . . . Soft showers
DR . . . Cascade showers
G . . Radiofrequency cosmic rays
H . . Solar cosmic rays
. . . * See also Cosmic ray photons BNG ORD C
K . *Source defined by a given radiation or particle*
. . * Use this position only when specifying a given particle.
. . * Add to BND K letters A/Y following BK;
. . * Add to BND L letters following BL;
. . * Add to BND letters N/Q following B; eg
KJ . . Ionizing radiation (source of particles)
LXT . . Gamma rays (source of particles)
. . . * Eg Neutrons - (By source) - Gamma rays BNW RDL XT.
OXH D . . Deuterons (source of particles)
. . . * Eg Neutrons - (By source) - Deuterons BNW ROX C.
By aggregation
RC . Cluster aggregates (particles)
RD . Many-particles systems
RE . . Micelles
RF . . Few-particles systems
By lifetime, mean life
RH . Stable particles
. . * Photons, leptons, mesons, baryons.

Elementary particles (types)

[Physics B]
[Matter BLY]
[Particles BM]
[Types of particles]
[Elementary particles types BNB]
[By lifetime]
. [Stable particles BNDRH]

BND RJ . Unstable particles
. . * With finite life, however long.
SL . . Long-lived particles
SN . . Short-lived particles
T . . . Resonances
. . . . * With greatly increased probability of interaction with colliding particles at the resonance energy.
. . . . * For psi particles (J particles), see BNS W.
TMD . . . . Quantum number properties
TMR U . . . . . Charm (resonances)
TU . . . . B resonances
V . . . Virtual particles, virtual quantum
. . . . * The uncertainty principle allows particles to be created for short periods in apparent violation of the energy conservation laws; eg pairs of virtual electrons and positrons in a complete vacuum.
By role in interaction
. * For exchange particles (exchange forces), see BMO.
X . Material particles (general)
. . * Ultimate constituents of matter; usually assumed to be point-like, with 1/2 integral spin.
. . * Do not use to specify particular particles.
By relation between quantum numbers
BNF . Antiparticles, antimatter, conjugate particles
. . * Retain mass and spin of their image particle, but all other quantum values are reversed. All particles except photons and pi-mesons have their anti-particle.
. . * See also Annihilation BMF V
By charge
BNF X . Electromagnetic field particles & waves, electromagnetic forces (particle physics)
. . * For charged particles, see BNH.
BNG . . Electromagnetic interaction (particles)
. . . Quantum mechanics
BNG 8M . . . . Quantum electrodynamics, QED
. . . . . Matrices
8M2 M . . . . . . Feynmann diagrams

[Elementary particles types BNB]
[Electromagnetic field particles & waves BNFX]
[Electromagnetic interaction particles BNG]
. [Quantum mechanics]
. . [Quantum electrodynamics BNG8M]
. . . [Matrices]
. . . . [Feynmann diagrams BNG8M2M]

. Scattering
BNG FTS . . Coulomb scattering, electrostatic scattering
FTT . . Rutherford scattering
. Ionizing radiation
KJ . . Nuclear radiation
. Exchange particles
O . . Photons
. . . * For gamma rays (high energy photons), see BLY.
O36 . . . Practical & experimental
. . . . * See also applications of photon interaction; eg Photon correlation spectroscopy B7M NGO.
. . . Scattering
OFT . . . . Compton scattering
OMC P . . . Supersymmetry
OMC Q . . . . Photinos
OQ . . . *Interactions with other particles*
OQG O . . . . Photon-photon interaction
. . . *Types of photons by source*
ORD C . . . . Cosmic ray photons
. . . . . * For cosmic ray X-rays, see X-rays BLX Q.
ORD H . . . . . Solar cosmic ray photons
ORL F . . . . High energy photons
BNH Electrically charged particles, charged particles
. * See also Alpha particles (nuclei) BOX HH; Electrons BNP; Positrons BNP RF; Protons BNV; Ions BQU; Cerenkhov radiation BNL N.
BNH U . Positive particles
V . Negative particles
W Zero charge particles, neutral particles
. Recombination
WFU M . . Neutralization
BNI Magnetically charged particles, magnetic monopoles, magnetic particles
. * Hypothetical particles, analogous to the electrical electron and proton, with north and south magnetic charges.
. Magnetic relaxation
BNI JBL R . . Spin-lattice relaxation
MM . Quantum number properties
. . Magnetic moment
MMN T . . . Gyromagnetic ratio
. . . . * Ratio of magnetic moment to angular momentum.
. . . . * See also Nuclear magnetic resonance BMO NO

Elementary particles (types)

[Particles BM]
[Elementary particles types BNB]
[By charge]
. . [Magnetically charged particles BNI]
. . . [Quantum number properties BNIMM]
. . . . [Magnetic moment]
. . . . . [Gyromagnetic ratio BNIMMNT]

By spin
BNJ . Fermions (general)
. . * Have 1/2 integral spin and observe Fermi-Dirac statistics. Comprise Leptons (BNM) & Baryons (BNT); all elementary particles except bosons.
BNJ MCP . . Supersymmetry
MCQ . . . Fermion partners of bosons
. . . . * The names of these are formed by changing the terminal -on of the boson to -ino; eg photino BNG OMC Q
BNK . Bosons
. . * All elementary particles except Fermions. Have integral spin and observe Bose-Einstein statistics.
. . * For gauge bosons, see Exchange particles BMO.
. . * See also Exchange particles BMP JO
BNK MCP . . Supersymmetry
MCQ . . . Boson partners of fermions
. . . . * The names of these are formed by prefacing the name of the fermion by s-; eg squarks BNR MCQ.
PJ . . Electroweak interactions
T . . Axions
. . . * Hypothetical light bosons, postulated to explain features of quantum chromodynamics (BNR 8M).
By energy characteristics
BNL C . Monoenergetic particles
D . Low energy particles
. . * <1 GeV.
E . Intermediate energy particles
F . High energy particles
. . * >1 GeV.
. . * For cosmic rays, see BND C.
G . Very high energy particles
H . Excited particles
. . * With more energy than in ground state.
K . Linear energy transfer particles
. . * See also Transfer reactions (nucleon-nucleon) BOQ UNR
L . . High linear energy transfer particles
M . Tachyons
. . * Postulated to have velocity exceeding that of electromagnetic waves.
. . * See also cosmic rays BND C
N . . Cerenkov radiation, Cherenkhov radiation
By mass, & strength of interaction shown
Q . Light particles (general)
R . Medium heavy particles (general)
S . Heavy particles (general)
. . * Heavier than pi-mesons.

[Physics B]
[Matter BLY]
[Particles BM]
[Types of particles]
[Elementary particles types BNB]
[By mass]
[Heavy particles general BNLS]

. * See also specific types; e.g. muons BNN; baryons BNT.
BNL U Generations (particles)
. * Families of particles.
W . Lepton & quark systems (generations)
BNM Leptons
. * Fermions which do not participate in strong interactions.
BNM PJ . Electroweak interactions
PJO . . W-bosons
QM . Lepton-lepton interaction
RLR . Heavy leptons (general)
. . * For muons, see BNN; for heavier-than-muon leptons, see BNN V.
BNN . Muons, mu-mesons
. . By source
BNN RDC . . . Cosmic ray muons
. . *By charge*
RHU . . . Positive muons
RHV . . . Negative muons
U . . Muonium
. . . * See also Positronium BNP RFT
V . Heavier than muon leptons
X . . Tauons, tau particles
BNO . Neutrinos
. . * Thought to have zero mass.
. . Antiparticles
BNO RF . . . Antineutrinos
. . *By associated particle*
RN . . . Muon neutrinos
RNR F . . . . Muon antineutrinos
RNX . . . Tauon neutrinos
BNP . Electrons
. . * See also Beta radiation BNP RDO; Beta decay BOF KR
BNP 36 . . Practical & experimental
. . . Separation
73P . . . . Millikan separation (electrons)
. . Theory
8T . . . Dirac electron theory
BD . . Energy levels, energy states
. . . * For electron energy states in particular media, see medium; eg in condensed matter BTX BD.
. . . * See also Photons BNG O
BEK . . . Excited state

[Types of particles]
[Elementary particles types BNB]
[By mass]
[Leptons BNM]
[Electrons BNP]
[Energy levels BNPBD]
. [Excited state BNPBEK]

	Emission
	. * For photoelectric effect, see BIV (general) and BVI V (solids).
BNP FG	. Electron field emission, autoemission, cold emission-
FGT	. . Secondary emission
	Collision
FS	. Electron impact, electron beam interaction, electron ionization
	. . * See also Atomic electron impact excitation BPF IXP; Molecular electron impact excitation BQF IXP
	Scattering
FT	. Compton scattering, Compton effect
FUS	Electron attachment, electron interaction
	. * See also radioactivity, BOF K
GB	Electron beams
	. Instruments
GB4 FV	. . Electron lenses, magnetic lenses
JR	Paramagnetism
	. * For electron paramagnetic resonance, see BNP MKO.
MK	Spin
MKF O	. Electron spin resonance, ESR, electron paramagnetic resonance, paramagnetic resonance
	. . * For electron spin resonance spectroscopy, see AY7 MKX.
MKF P	. Electron spin polarization
	. Magnetic relaxation
MKJ CP	. . Electron spin-lattice relaxation
	Types of electrons
	. * For orbital electrons, see Atomes BPD T.
	. *By source*
RDC	. . Cosmic ray electrons
RDK J	. . Delta radiation
	. . . * Secondary electrons (or protons) emitted as a result of ionizing radiation.
RDO	. . Beta particles, beta radiation
	. . . * For beta decay, see BOF KS.
RDQ	. . Normal electrons
	. *Aggregations, generations*
RDR D	. . Many-electron systems
RDR F	. . Electron pairs
	. Anti-particles
RF	. . Positrons, positons, positive electrons

[Elementary particles types BNB]
[By mass]
[Leptons BNM]
. [Electrons BNP]
. . [Types of electrons]
. . . [Anti-particles]
. . . . [Positrons BNPRF]

	. . . . . Positron annihilation
BNP RFF V	. . . . . . Electron pair annihilation
RFQ P	. . . . . Electron-positron interaction
	. . . . . . * For Electron pair annihilation see BNP RDR FFV.
RFQ PFS	. . . . . . Electron-positron collision
	. . . . * Producing hadrons.
RFT	. . . . . Positronium
	. . . . . . * Short-lived association between positron and electron.
	. . . . . . * See also Muonium BNN U
RFU	. . . . . . Orthopositronium
RFV	. . . . . . Parapositronium
	. . . *By charge*
RHV	. . . . Normal negatively charged electrons
	. . . *By energy*
RLD	. . . . Low energy electrons
RLF	. . . . High energy electrons
RLJ	. . . . Free electrons
	. . . . . * See also conduction electrons (semi-conductors) BVI FN
	. . . *By associated particle*
RNO	. . . . Electron neutrinos
RNO RF	. . . . Electron antineutrinos
T	. . . Inner shell electrons
V	. . . Valency electrons
W	. . . Hot electrons
BNQ	Hadrons (general)
	. * All elementary particles except leptons and photons.
BNQ 8B	. Theories, models
8SC	. . Composite hadron models
8SL	. . Bootstrap models
8SN	. . Duality models
8SP	. . Peripheral models
8SP Q	. . . Multiperipheral models
8W	. . Classification models (hadrons)
	. . . * For Parton model, see BNU RBW; for Quark model, see BNR.
	. *Properties & processes*
	. . Very high energy
BCI	. . . Hadron-induced very high energy interactions
BD	. . Energy levels
BEM	. . . Bound state
FS	. . Collision
FSH	. . . Cross section
FT	. . Scattering
	. . . Matrix algebra
FT2 TB	. . . . S-matrix

[Elementary particles types BNB]
[By mass]
[Hadrons general BNQ]
[Properties & processes]
[Scattering BNQFT]
. [Matrix algebra]
. . [S-matrix BNQFT2TB]

	. Models
	. . Regge poles
BNQ FT8 SR	. . . Reggeons
	. Inelastic scattering
FTJ N	. . Hadron induced inelastic scattering
FTJ O	. . . Two-particle final states (hadron scattering)
FTJ P	. . . Many particle final states (hadron scattering)
HK	Charge
	. * See quantum properties BNQ MM
M	*Quantum number properties*
MB	. Symmetry
MCS	. . Unitary symmetry, internal symmetry
	. . . * For the hypothetical groups characterized by this property, see multiplets BNQ T.
	. . . * See also Isospin BNQ MI
	. Parity
MEK	. . Charge conjugation parity, C-parity
MH	. Mass
MI	. Isospin, isotopic spin
MIR	. . I (isospin)
MIS	. . I3 (isospin)
MIT	. . . SU2 (isospin)
MJ	. Angular momentum
MM	. Charge
MMT	. . Hypercharge
MPT	. . Baryon number
	. Strangeness
MRW	. . Strange particles (hadrons)
	. . . * See also Hypercharge BMM MT
PJ	Electroweak interactions
	. * For electromagnetic interactions, see BNG.
	. Weak interactions
PL	. . Weak hadron interactions
PN	Strong interactions (general)
	. * Extremely short range force, particularly that holding protons and nueutrons together in the atomic nucleus. Primarily the result of the exchange of gluons between quarks, the constituents of hadrons; see BNR O.
PNO	. Exchange particles
	. . * For gluons, see under quarks BNR O.
Q	*Interactions with other particles*
QGO	. Photon-hadron interaction
QGO FT	. . Scattering
QH	. Charged hadron interaction
QM	. Lepton-hadron interaction
QMP L	. . Weak lepton-hadron interaction
QMQ H	. . Charged lepton-hadron interaction

[Elementary particles types BNB]
[By mass]
[Hadrons general BNQ]
[Properties & processes]
. [Interactions with other particles BNQQ]
. . [Lepton-hadron interaction BNQQM]
. . . [Charged lepton-hadron interaction BNQQMQH]

BNQ QN	. . . Muon-hadron interaction
QP	. . Electron-hadron interaction
QPF T	. . . Scattering
QPF TJL	. . . . Elastic scattering
QPF TT	. . . . Compton scattering
QQ	. . Hadron-hadron interaction
QYO	. . Nucleus-hadron interaction
	. . . * See Nucleus BOQ Q
	Types of hadrons
	. *By part*
RBW	. . Composite models
	. *By special & quantum property*
RCB	. . Symmetrical
	. . . * For multiplets, see BNQ T.
	. *By lifetime*
DT	. . Resonances
T	. Multiplets, isospin multiplets
	. . * Hypothetical groups of particles in which all the particles are regarded as different states of the same particle. All have the same spin (J), parity (P) & baryon number (B).
	. . * See also Hadron symmetry BNQ MC
U	. . SU3 (multiplets)
	. . . * Special unitary group of 3x3 matrices, predictive of hadronic multiplet structure.
	. . . Theory
U8U	. . . . Eightfold way (hadron theory), octet theory
BNR	. . . Quarks
	. . . . * Hypothetical particles corresponding to the basic building blocks of the SU3 group. Their charge is not the usual electron charge (e) but integral multiples of 1/3e.
	. . . . Theory
BNR 8M	. . . . . Quantum chromodynamics, QCD (quarks)
8MG	. . . . . . Quark confinement
	. . . . *Properties*
MCP	. . . . . Supersymmetry
	. . . . . . * See also Axions BNK T
MCQ	. . . . . . Squarks
MRS	. . . . . Colour (quarks)
	. . . . . . * See also the types of quarks below characterized by different manifestations of this property.
MRU	. . . . . Charm
	. . . . Exchange particles
O	. . . . . Gluons
	. . . . *Types of quarks*
RF	. . . . . Antiquarks

Hadrons (general)

[Hadrons general BNQ]
[Types of hadrons]
[Multiplets BNQT]
. [SU3 multiplets BNQU]
. . [Quarks BNR]
. . . [Types of quarks]
. . . . [Antiquarks BNRRF]

BNR SF . . . . Flavours
. . . . *With 2/3 the proton charge*
SP . . . . . Up (quark flavours)
SU . . . . . Charm (quark flavours)
T . . . . Top (quark flavours)
. . . . *With 1/3 the proton charge*
TX . . . . . Down (quark flavours)
U . . . . . Strange (quark flavours)
UX . . . . . Bottom (quark flavours)
VF . SU4 (multiplets)
VH . SU6 (multiplets)
BNS Mesons
. * Bosons, with mass usually intermediate between
. electrons and nucleons.
. * For mu-mesons, see under Leptons BNN.
. Interactions
BNS FJ . . Decay
FJM . . . Leptonic decay (mesons), semi-leptonic
. . . decay
FJM P . . . . Pi leptonic decay
FJM T . . . . K leptonic decay
FJQ . . . Hadronic decay (mesons)
. *Properties*
MEG . . G-parity (mesons)
Q . . *Interactions with other particles*
. . . * For Meson-nucleon interactions, see Nucleons
. . . BNU QS; for Meson-hyperon interactions, see
. . . Hyperons BNX QS.
QS . . . Meson-meson interaction
. *Types*
. . *By property*
RCR U . . . Charmed mesons
. . *By source*
RDC . . . Cosmic ray mesons
. . *By time*
. . . Resonances
RDT . . . . Meson resonances, vector resonances
. . . . (mesons)
RDU . . . . . A-resonances (mesons)
RDV . . . . . B-resonances (mesons)
. . *By mass*
RLQ . . . Light mesons
RLR . . . Heavy mesons
T . . Pi-mesons, pions
TFJ . . . Decay
TQ . . . *Interactions with other particles*
. . . . * For pion-baryon interactions, see Baryons
. . . . BNT QST.
TRH . . . Charged pi-mesons
TRH U . . . . Positive

[Hadrons general BNQ]
[Types of hadrons]
[Mesons BNS]
. [Types]
. . [Pi-mesons BNST]
. . . [Charged pi-mesons BNSTRH]
. . . . [Positive BNSTRHU]

BNS TRH V . . . . Negative
U . . K-mesons, kappa mesons, kaons
UFJ . . . Decay
UFJ Q . . . . Hadronic decay (kaons)
URH . . . Charged kaons
URH W . . . Zero charged kaons
V . . Eta-mesons
VFJ . . . Decay
VPL . . . Weak interactions
VPN . . . Strong interactions
W . . . J particles, psi particles, psi mesons,
. . . psi resonances
WMB . . . . Symmetry
WMR S . . . . Colour
WMR U . . . . Charm
WQ . . . . *Interactions with other particles*
WQQ T . . . . . Multiplets
WQQ U SU3
WQR . . . . Quarks
WQR MM . . . . Charge
Y . . Other mesons [
. . . * Arrange A/Z by equivalent of Greek initial
. . . letters; eg Upsilon meson BNS YU; Rho
. . . meson BBS YP
BNT Baryons
. * Fermions which decay into nucleons by
. emisssion of mesons.
BNT FC . Propagation
FCS . . Photoproduction of baryons
FJ . Decay
FJM . . Leptonic decay (baryons)
FJQ . . Hadronic decay (baryons)
. *Interactions with other particles*
. . * For baryon-hadron interaction, see Hadronic
. . decay BNT FJQ Q
QM . . Baryon-lepton interaction
QST . . Pion-baryon interaction
QSU . . Kaon-baryon interaction
. *Types*
RDT . . Baryon resonances
RF . . Antibaryons
RH . . Charged baryons
T . . Dibaryons
BNU . Nucleons
. . Motions
BNU DKT . . . Collective motions (nucleons)
. . . . Transitions
DKT BDN . . . . Collective transitions
MK . . Spin
. . . Exchange of spin direction

Hadrons (general)

[By mass]
[Hadrons general BNQ]
[Types of hadrons]
[Baryons BNT]
[Nucleons BNU]
[Spin BNUMK]
. [Exchange of spin direction]

BNU MKN X . . Bartlett force
MW Composite models
. * For parton model, see BNU RBP.
Interactions with other particles
QU . Nucleon-nucleon interactions
. . * See also Nuclear forces BOB G
QUF L . . Absorption
QUF S . . Collisions
QUF T . . Scatter
QUM NX . . Exchange forces
QUM NYH . . . Heisenberg force
QUM NYM . . . Majorana force
Types
RBW . Partons
. . * See also Quarks BNR
. . Interactions with other particles
RBW QST . . . Parton-pion interaction
. *By aggregation*
RDR F . . Few-nucleon systems
RF . Antinucleons
BNV Protons, protonium
BNV 7T3 U . Accelerators
7VC . . Synchrotons
7VD . . . Zero gradient synchrotons
7VG . . . Alternating gradient synchrotons
7VL . . . . Fixed field alternating gradient synchrotons, ring accelerators
. *Interactions with other particles*
QU . . Proton-nucleon interaction
QUR BW . . Parton-proton interaction
QUR BWF T . . . Scatter
QV . . Proton-proton interaction
QVW . . . Proton-proton inclusive interaction
. *Types*
. . * For delta radiation, see BNP RDK J.
RDC . . Cosmic ray protons
RF . . Anti-protons
T . . Delayed protons
BNW Neutrons
BNW AT . Transport
. Interactions
. . Flux
CW . . . Neutron flux, neutron economy
CWC L . . . . Neutron flux density
FJ . . Decay
FKS . . . Beta decay
FQ . . Diffraction
FUF . . Diffusion

[Hadrons general BNQ]
[Types of hadrons]
[Baryons BNT]
[Nucleons BNU]
. [Neutrons BNW]
. . [Interactions]
. . . [Diffusion BNWFUF]

BNW FUH . . . . Multigroup diffusion
. . . *Interactions with other particles*
QW . . . . Neutron-neutron interactions
. . *Types of neutrons*
. . . *By non-quantum properties*
. . . . Beta decay
RBF KS . . . . . Delayed neutrons
. . . *By source*
RDA . . . . Particle accelerators neutron source
RDA R . . . . Nuclear reactors neutron source
RDC . . . . Cosmic ray neutrons
RDL XT . . . . Gamma radiation neutrons
RDN V . . . . Protons (neutron source)
RDO XHD . . . . Deuterons (neutron source)
RDT . . . Resonances
RDU D . . . . Dineutrons
RDU F . . . . Tetraneutrons
RF . . . Anti-neutrons
. . . *By energy*
RLC . . . . Monoenergetic neutrons
RLD . . . . Thermal neutrons, slow neutrons
RLD N . . . . Cold neutrons
. . . . . * Very low energy neutrons, from reactor.
RLE . . . . Intermediate neutrons
RLF . . . . Fast neutrons
. . . . . * >1MeV.
BNX Hyperons
. * All long-lived baryons other than nucleons.
. Interactions with other particles
. . * For hyperon-nucleus interactions, see BOQ NX.
BNX QNS . . Meson-hyperon interaction
. *Types*
RF . . Anti-hyperons
T . . Lambda particles
. . . Interactions with other particles
TPW . . . . Lambda-neutron interactions

Nuclei

[Particles BM]
[Types of particles]
. . . [Hadrons general BNQ]
. . . . [Types]
. . . . [Lambda particles BNXT]
. . . . [Interactions with other particles]
. . . . [Lambda-neutron interactions BNXTPW]

. . . . * For hypernuclei, see
. . . . BOX FT.
. . . . Anti-particles
BNX TRF . . . . Antimatter state
. . . . (hyperons)
U . . . . Sigma particles
URF . . . . Antiparticles
URF RH . . . . Charged
URF RHN . . . . Zero charged
URH . . . . Charged
URH W . . . . Zero charged
V . . . . Xi particles, cascade particles
VRF . . . . Antiparticles
VRH . . . . Charged
VRH W . . . . Zero charged
W . . . . Omega particles, omega-
. . . . minus particles
. . . . * See also Unitary symmetry
. . . . BMM CS
X . . . . Other hyperons
. . . . * Arrange A/Z.
BNY Atomic & nuclear physics (together)
BO . Nuclei, nucleus, nuclear physics
. . * Works on nuclear physics using the term in
. . its very broad sense (to include elementary
. . particles) should go at BM.
. . * Note that the instructions at BNB for
. . qualifying and specifying any given particle,
. . using letters A/S, are adjusted for nuclei
. . (from BOQ YC onwards) in order to
. . accommodate the special extension of the
. . reactions facet. So Types of nuclei begin at
. . BOX D, not BOR.
BO3 6 . . Practical & experimental physics
. . . Plant
W . . . . Nuclear reactors
BO7 H . . . Track visualization
IJ . . . . Particle beam handling
T . . . Acceleration
TN . . . . *Types of accelerator*
. . *General properties*
BO8 B . . Theory & models
M . . . Quantum theory
M2M 8G . . . . Perturbation theory
M2M 8GE . . . . . Born approximation
P . . . . Wave mechanics
P2M 9NS . . . . . Schrodinger's equation
VC . . . Unified model (nucleus)
VE . . . Nuclear scattering model

[Physics B]
[Matter BLY]
[Particles BM]
[Atomic & nuclear physics together BNY]
[Nuclei BO]
[Theory & models BO8B]
. [Nuclear scattering model BO8VE]

BO8 VF . . Optical model (nucleus)
VH . Nuclear clustering model, alpha-
. particle model (nucleus)
VJ . Nuclear collective model
VK . . Nuclear cranking model
VL . . Liquid drop model (nuclear), drop
. . model (nuclear)
VN . Nuclear shell model, independent
. particle model (nucleus)
VP . Hartree-Fock model
VR . Other models
. . * Arrange A/Z; e.g. Resonating group
. . structure model BO8 VRR.
Processes & properties
BO9 D . Spatial dimensions
. . * For size of nucleus as a quantum
. . number property, see BOM G.
G . Systems characteristics
GV . . Conservation
J . . Structure
. . . * See also nuclear models BO8 B
JS . Symmetry
. . * Use BOM B
. Configuration
JV . . Shape of nucleus
JVH . . . Deformation (nuclear)
JVS . . . Surfaces (nucleus)

BOA F Energy interactions & forms
. * For nuclear reactions in the narrow sense,
. see BOR.
BOB . Mechanics
BOB B . . Nuclear energy (general)
. . . * For Nuclear reaction energy, see
. . . BOR BB
D . . . Nuclear energy levels
. . . . * For binding energy, see Nuclear
. . . . forces BOB GH.
D9C I . . . . Nuclear energy levels lifetime
D9D J . . . . Energy level width
. . . . Transitions nuclear energy
. . . . levels
DN . . . . . Nuclear electromagnetic
. . . . . transition
. . . . . . * Change from one quantum
. . . . . . energy level to another.
DNF . . . . . Internal conversion (nuclear
. . . . . transitions)

[Nuclei BO]
[Energy interactions & forms BOAF]
[Mechanics BOB]
[Nuclear energy general BOBB]
. [Nuclear energy levels BOBD]
. . [Transitions nuclear energy levels]
. . . [Internal conversion nuclear transitions BOBDNF]

BOB DNG . . . Mossbauer effect
DNJ . . . Branching & mixing ratios (nuclear
. . . transitions)
DQ . . Nuclear isomerism
. . . * Existence of different energy states in
. . . otherwise identical nuclides and resulting in
. . . different radioactivity characteristics.
DQD N . . . Transitions
DQD O . . . . Isomeric transitions
DQD P . . . . . Independent decay
DQD R . . . . . Stable isomers
DR . . Isobaric analogue state
DS . . Nuclear collective state
DW . . Nuclear orientation, alignment
Force
G . Nuclear forces
. . * For nuclear exchange forces, see BON X.
GH . . Binding energy
GJ . . . Mass defect
GK . . . Coulomb energy (nuclear forces)
GN . . Nucleon-nucleon forces
. . . * For charges, see BOM M.
. . . * See also nucleon-nucleon interaction BNU QU
GN8 S . . . Meson field theory
GS . . . Central force
GT . . . Tensor force
GV . . . Spin-orbital coupling
IM . Moments
BOC ER . Repulsive force (nuclear)
H Statics
J . Nuclear mass
. . * For mass as a quantum number property, see
. . BOM H.
L . Nuclear density
P . Nuclear stability

[Particles BM]
[Atomic & nuclear physics together BNY]
[Nuclei BO]
[Energy interactions & forms BOAF]
[Mechanics BOB]
[Statics BOCH]
. [Nuclear stability BOCP]

BOC X Dynamics
BOD E . Kinetics
EN . . Coupled channels
N . Rotation
. . * For angular momentum, see quantum
. . properties BOM J; for spin, see
. . BOM K.
BOF . Wave motion
BOF G . . Emission nuclear reactions
GR . . . Prompt emission processes
GS . . . Delayed emission processes
IN . . . Excitation
IV . . . . Cascade reactions
. . Decays
J . . . Disintegrative processes,
. . . spontaneous reactions
. . . . * For nuclear reactions resulting from
. . . . the bombardment of nuclei by
. . . . particles, see BOR. Use the latter
. . . . also for works dealing with nuclear
. . . . reactions in general.
J8B . . . . Nuclear decay theory
JFC N . . . . Damping
JN . . . . *By radiation/particle emitted*
. . . . . * Add to BOF JN letters A/Y
. . . . . following BN; eg
JNW . . . . . Neutron product reaction
JP . . . . Spallation
. . . . . * Very vigorous decay, usually
. . . . . following bombardment.
JT . . . . Transmutation
K . . . . Radioactivity
. . . . . * Spontaneous decay of heavy
. . . . . elements.
K9C N . . . . . Half-life, lifetime
KQW . . . . . Atmospheric radioactivity
. . . . . . * Note preceding BND A applies
. . . . . . here also.
. . . . . . * See also DS Atmosphere
KQX . . . . . . Fallout
KR . . . . . Alpha radioactivity, alpha
. . . . . decay, alpha radiation
KS . . . . . Beta radioactivity, beta decay
KT . . . . . . Electron capture, electron
. . . . . . attachment
. . . . * Transformation of the
. . . . nuclide in which a bound
. . . . electron

[Dynamics BOCX]
[Wave motion BOF]
[Decays]
. [Disintegrative processes BOFJ]
. . [Radioactivity BOFK]
. . . [Beta radioactivity BOFKS]
. . . . [Electron capture BOFKT]

. . . . . merges with the nucleus.
BOF KU . . . . . Radiative capture, orbital electron
. . . . . capture
KV . . . . . . K-coupling (beta decay)
KW . . . . . . L-capture
KX . . . Gamma radioactivity
O Resonance
. * See also magnetic effects BOJ FO
OQ . Giant resonances
OR . Isobaric analogue resonance
OS . Yield (nuclear reactions)
P Nuclear polarization
. * See also nuclear orientation BOB DW
RX Inclusive interactions
Collisions
S . Bombardment (nuclear reactions)
. . * See also Compound nuclear processes BOF SR.
S8B . . Theory
S8S E . . . Many-body theory
S8W L . . . Lane-Robson theory
S8W N . . . Yang theory
SBB . . Energy
SBF N . . . Excitation functions
SBF T . . . Threshold energy
SCU . . Momentum
. . . * For angular momentum, see quantum number
. . . properties, BOM J.
SCU R . . . Transverse momentum (nuclear)
SCU U . . . Longitudinal momentum (nuclear)
SH . . Cross sections
SH9 DJ . . . Mean free path
. . *Particular forms of collisions*
SI . . . Impact
. . . . * With high rate of change of momentum,
. . . . generating large contact forces.
SL . . . Elastic collision
SM . . . Inelastic collision
SN . . . . Collisions of the first kind
SO . . . . Collisions of the second kind
SP . . . One dimensional collision
SQ . . . Binary collision
SR . . . Compound nuclear processes
. . . . * Forming highly excited and short-lived
. . . . nuclei immediately after collisions.
T Scattering
. * In narrow sense: energy is transferred, but without
. removal or addition of nucleons.
. *Special quantum models*
T8S J . . Form factors
TJL . Elastic scattering

[Nuclei BO]
[Energy interactions & forms BOAF]
[Mechanics BOB]
. [Dynamics BOCX]
. . [Wave motion BOF]
. . . [Scattering BOFT]
. . . . [Elastic scattering BOFTJL]

. . . . . * For Coulomb scattering, see BNG FTS.
BOF TJM . . . . Quasi-elastic scattering
TJN . . . . Inelastic scattering
. . . . *Special to nuclei*
TT . . . . . Shadow scattering
TU . . . . . Few neutron scattering
UR . . . Capture
. . . . * For the capture of specific particles, see
. . . . nuclear reactions; eg nuclear muon capture
. . . . BOQ NFU R.
. . . . * See also Direct nuclear reactions BOR W;
. . . . Electron attachment (radioactivity) BOF KT
US . . . . Stripping (nuclear)
. . . . . * A part of the incident nucleus merges with
. . . . . target nucleus, the remainder proceeding
. . . . . largely unchanged.
USP . . . . . Oppenheimer-Phillips process, OP
. . . . . process (nuclear)
UT . . . . Pick-up reaction
BOH Electro-magnetic phenomena
BOH BIM . Electromagnetic moment
I . Electricity
IBI M . . Electric moment
. . . * See Charge BOM MP
K . Charge
. . * See quantum number properties, BOM M
BOJ . Magnetism
. . * For nuclear magnetic moment, see BOM NJ
BOJ FO . . Resonance
. . . * For nuclear magnetic resonance, see BOM NO.
BOL Y Nuclear matter
. * For nuclear energy in general, see BOB B.

BOM Quantum number properties
BOM B . Symmetry
E . Nuclear parity
FM . Mean life
. . * For lifetime in radioactivity, see BOF KQF.
G . Size of nucleus
GJ . . Width
GK . . Radius
H . Mass
HP . . Packing fraction
I . Isospin (nuclei), isotopic spin (nuclei)
. . * A quantum mechanical variable.
. . * See also Hadrons BN
J . Angular momentum
. Spin
K . . Nuclear spin
KR . . . Helicity

[Atomic & nuclear physics together BNY]
[Nuclei BO]
[Energy interactions & forms BOAF]
[Quantum number properties BOM]
. [Spin]
. . [Nuclear spin BOMK]
. . . [Helicity BOMKR]

BOM KS . . . Orbital spin
. . . . * For spin orbital coupling, see Nucleons
. . . . BOB GV.
KV . . . Intrinsic spin
M . Charge
M92 D . . Charge distribution
MB . . Charge symmetry
MO . . Charge independence
. . Electric moment
MP . . . Quadrupole moments (nuclear)
MT . . Hypercharge
. *Magnetic properties*
NJ . . Nuclear magnetic moment
NK . . . Dipole magnetic moment
NL . . . Quadrupole magnetic moment
NM . . . Octopole magnetic moment
O . . . Magnetic resonance (nuclear), NMR
. . . (general)
. . . . * Most of the literature relates to its
. . . . application in spectroscopy, etc (see
. . . . B7M OO).
OP . . . . Double nuclear magnetic resonance
OQ . . . . . Internuclear nuclear magnetic
. . . . . resonance, INDOR
OQV . . . . . . Overhausen effect
OR . . . . Quadruple resonance (nuclear)
PM . Lepton number
PT . Baryon number
RU . Charm
RW . Strangeness

Interactions by energy expenditure
BON J . Superhigh energy reactions
V Basic interactions
X . Nuclear exchange forces
BOQ *Interactions with other particles*
. * This class (BOQ) takes general studies of the
. interactions betweeen a particle and a nucleus,
. without specification of the role(s) played by each.
. * Reactions in which roles are specified are dealt with
. in classes BOR/BOX
. * Add (retroactively) to BOQ letters B/X following
. BN; eg Hyperon-nucleus reactions BOQ X.
. * Add (retroactively) to BOQ Y letters O/Q
. following B; eg Nucleus-nucleus reactions BOQ YO.
BOQ DC . Cosmic rays-nucleus reactions
DCF T . . Scattering
DLX T . Gamma radiation-nucleus reactions
GO . Photon-nucleus reactions

[Matter BLY]
[Particles BM]
[Atomic & nuclear physics together BNY]
[Nuclei BO]
[Energy interactions & forms BOAF]
[Interactions with other particles BOQ]
. [Photon-nucleus reactions BOQGO]

. . High energy photon reactions
BOQ GOR LF . . . Photodisintegration
M . Lepton-nucleus reactions
N . . Muon-nucleus reactions
NFU R . . . Capture
NX . . Hyperon-nucleus reactions
P . . Electron-nucleus reactions
. . . * For nuclear-electron capture, see
. . . Beta decay BOF KT.
Q . Hadron-nucleus reactions
QFT . . Scattering
S . . Meson-nucleus reactions
U . . Nucleon-nuclear reactions
. . . *Special reactions*
UNR . . . . Transfer reaction
. . . . . * Nucleons are exchanged between
. . . . . target nucleus and the projectile
. . . . . nucleus.
. . . . . * See also Linear energy transfer,
. . . . . BNL K
URD RF . . . Few nucleons-nuclear reactions
W . . . Neutron-nucleus reactions
. . . . Capture
WFU R . . . . . Neutron radiative capture
X . . Hyperon-nucleus reactions
XFU R . . . Hyperon capture
YO . *Nucleus-nucleus reactions*

BOR Nuclear reactions (general)
. * Reactions between nucleus & bombarding
. particles.
. * The instructions at BNB for the
. subdivision of any given particle are
. amended here; see second note at BO
. above.
. * Class here (BOR) general works on all
. nuclear reactions, including spontaneous
. ones. For spontaneous reactions alone, see
. BOF J.
. * Reactions in which roles are specified are
. dealt with in classes BOS/BOX (see the
. instructions preceding BOS).
. * See also radiation and wave phenomena
. (BOF/BOG) for constituents of reactions
BOR 2M . Mathematics
2SA . . Group theory (nuclear reactions)
2SA G . . . G-matrix
2SA K . . . K-matric
2SA R . . . R-matrix
. Theory & models

Nuclear reactions (general)

[Matter BLY]
[Particles BM]
[Atomic & nuclear physics together BNY]
[Nuclei BO]
[Energy interactions & forms BOAF]
[Nuclear reactions general BOR]
. [Theory & models]

BOR 8VL . . Optical model
. Energy
BB . . Nuclear reaction energy, Q, Q-factor, Q-value
BB9 2D . . . Energy distribution
. . . . * Between target amd interacting particles.
BCJ . . . Mass-energy relation
. *By general characteristics*
. . * For entoergic and exoergic nuclear reactions, see Collisions BOF SM.
PN . . Natural nuclear reactions
. . . * For spontaneous decay, see BOF K.
PR . . Artificial nuclear reactions
Q . . Chain reactions
. . . * For cascade reactions, see Excitation BOF IV.
QT . . Knock-on nuclear reactions
R . . Cyclic nuclear reactions
S . . Rearrangement processes (nuclear reactions)
V . . Exchange reactions
. . . * For transfer reaction, see BOQ UNR.
W . . Direct nuclear interactions
. . . * See also capture reactions BOF UR
X8V . . . Jackson model
Nuclear reactions by roles played
. * Reactions in which roles are specified observe the following citation order (which can also be seen in the inverted filing order of the arrays concerned):
. * (1) By target nuclide (2) By incident & emitted particle or radiation combined (3) By projectile, incident particle (4) By particle emitted (5) By product nucleus. For example: Lithium as target, proton as projectile, with He4 as product BOV HLT VSH H.
BOS . *By product nucleus*
. . * Add to BOS letters D/R following BOX; eg
BOS HD . . Deuteron (product nucleus)
HJ . . Helium-4 (product nucleus)
. . . Scattering
HJF T . . . . Alpha particle-nucleus scattering
HL . . Lithium (product nucleus)
JC . . Magnesium (product nucleus)

[Particles BM]
[Atomic & nuclear physics together BNY]
[Nuclei BO]
[Energy interactions & forms BOAF]
[Nuclear reactions by roles played]
[By product nucleus BOS]
. [Magnesium product nucleus BOSJC]

By emitted radiation or particle
. * Add to BOS letters T/Y following BOT; eg
BOS T . Gamma rays (emitted particles)
V . Proton (emitted particle)
W . Neutron (emitted particle)
YR . Nuclear fragments (emissions)
BOT *By projectile, incident radiation or particle*
. *Nuclides as projectiles*
. . * Add to BOT letters D/R following BOX; eg
BOT HD . . Deuteron (projectile)
HE . . Triton (projectile)
HH . . Alpha particle (projectile in nuclear reactions)
. *Incident particles/radiation*
T . . Gamma rays (incident radiation)
U . . Meson (projectile)
V . . Proton (projectile)
W . . Neutron (projectile)
X . . *Other particles*
. . . * Add to BOT X letters B/Y following BN;
. . . * Add to BOT Y letters O/Q following B; eg
XGO . . Photon projectile, photonuclear reactions, photo-disintegration
YQU S . . Heavy ion projectile
BOU *By incident & emitted radiation/particle combined*
. * Add to BOU letters B/Y following BOT; eg
BOU V . Proton (incident & emitted particle combined)
. . * For an example, see under chlorine as target, bombarded by protons with the emission of 6 protons & 4 neutrons, and production of magnesium 28: BOV KLU VSW SJC.
By target nucleus
Y . Targets (nuclear bombardment)
. . * Add to BOV letters D/H following BOX; eg
BOV DDRN . . Radioactive nuclei
. . *By mass number*
GB . . . Nuclei with mass numbers 1-5
. . *By element*
HD . . . Deuteron (target nucleus)

[Nuclei BO]
[Energy interactions & forms BOAF]
[Nuclear reactions by roles played]
[By target nucleus]
. [Targets nuclear bombardment BOUY]
. . [By element]
. . . [Deuteron target nucleus BOVHD]

BOV HE . . . Triton (target nucleus)
HH . . . Helium (target nucleus), He4 (target nucleus), alpha particles(target nuclei)
HI . . . He3 (target nucleus)
HL . . . Lithium (target nucleus)
HLT V . . . . Bombarded by protons
HLT VSH H . . . . . He4 product
HLT VSH HBH L . . . . . . Mass defect
. . . *Other nuclides*
. . . . * Add to BOV letters I/N following C in Chemistry, except for thorium, uranium & plutonium (which are separately notated); eg
IU . . . . Beryllium (target nucleus)
JK . . . . Carbon (target nucleus)
KL . . . . Chlorine (target nucleus)
KLU V . . . . . *Bombarded by & emitting protons*
KLU VSW . . . . . . *Emitting neutrons also*
KLU VSW SJC . . . . *With production of magnesium*
NCF . . . Neptunium (target nucleus)
P . . . Thorium (target nucleus)
Q . . . Uranium (target nucleus)
R . . . Plutonium (target nucleus)
S . . . Trans-uranium nuclides (target nuclei)

By mode of energy release
BOW . Fission
. . * Disintegration of heavy nuclei into two or more lighter ones, with a loss of mass. Usually the result of the impact of a neutron.
. . * Add to BOW letters A/V following BO; eg
BOW 3W . . Fission reactors
. . Theory & models
8VL . . . Liquid drop model fission
BB . . Energy
FK . . Spontaneous fission
. . *By product particle*
S . . . Fission output
SV . . . . Protons (fission products)
SW . . . . Neutrons (fission products)
SW9 BG . . . . . Number of neutrons
SWT . . . . . Prompt neutrons
SWV . . . . . Delayed neutrons
SYR . . . . Fission fragments

[Nuclei BO]
[Nuclear reactions by roles played]
[By mode of energy release]
[Fission BOW]
. [By product particle]
. . [Fission output BOWS]
. . . [Fission fragments BOWSYR]

. *By projectile*
BOW TW . . Neutrons
. . Photon
TXG O . . . Photofission, nuclear photoeffect
. . . *By particle emitted*
TXG OST . . . . Gamma fission reactions
TXG OSV . . . . Proton fission reactions, photoproton reactions
TXG OSW . . . . Neutron fission reactions, photoneutron reactions
TYQ US . . Heavy ion
UY . *By target*
VP . . Thorium
VQ . . Uranium
VR . . Plutonium

BOX B Nuclear fusion, thermonuclear reactions
. * Nuclear reaction between two light atomic nuclei to form a single heavier nucleus of greater binding energy (usually helium) with the loss of mass producing nuclear energy.
. * Add to BOX B letters A/V following BO, with the adjustments indicated; eg
B3R . Materials
. . * For plasmas, see BOX B73.
B3W . Fusion reactors
. Production
. . * For plasma confinement, see BOX B7R.
B73 . . Plasmas (nuclear fusion)
. . . * For particular types of plasmas, see BOX B8V.
. . . * Add to BOX B73 letters K/P following BRV 73; eg Excitation BOX B73 M.
B7Q . Acceleration of particles
. . * This amends the notation at BM (where particle acceleration is BM7 T) in order to accommodate the vocabulary of plasma techniques below. The details from BM7 are available here if required (see Add instructions below).
. . * For acceleration of plasmas, see BOX B73 P.
. . * Add to BOX B7Q letters A/Q following BM7 T;
. . * Add to BOX B7Q S letters T/X following BM7 T (for types of accelerators);
. . * Add to BOX B7Q letters U/V following BM7 (for more types of accelerators).
. *Special techniques*
B7Q Y . . Thermonuclear systems (techniques)

Nuclei

[Nuclei BO]
[Energy interactions & forms BOAF]
[Nuclear reactions by roles played]
[By mode of energy release]
[Nuclear fusion BOXB]
[Special techniques]
[Thermonuclear systems techniques BOXB7QY]

BOX B7R	Plasma containment, confinement
	. *Properties, parameters*
B7R R	. . Lawson criterion
	. . . * Product of particle density and containment time.
B7R S	. . Containment time
B7R T	. . Confinement temperature
B7R U	. . Plasma density
B7R X	. Electrostatic cofinement
B7S	. Magnetic lines
B7S S	. . Internally generated magnetic lines
B7S T	. . Externally generated magnetic lines
B7S U	. . Open magnetic lines
B7S V	. . Closed magnetic lines
B7T	. Magnetic wells
	. . * Configuration of magnetic field for containment.
B7U	. . Linear field confinement, axial field confinement
B7U 3YB	. . . Magnetic bottles
B7U 3YD	. . . . Magnetic mirrors
B7V	. . Pinch effect (general)
B7V R	. . . Poloidal field confinement
B7V T	. . . Toroidal field confinement
B7V U	. . . . Zeta pinch
B7V V	. . . . Theta pinch
B7V W	. . . . Tokamak stellarator
B7W C	. . Cusp systems
B7W E	. . Baseball coils
B7W G	. Radiofrequency confinement
B7W K	. Non-magnetic confinement
B7W L	. Inertial confinement
B7W N	Rotating plasma systems
B7W P	Plasma injection systems
B7W R	Thermal systems (nuclear fusion)
B7W T	Fusion energy extraction
	Plasma reactions
	. * Normal retroactive synthesis is interrupted here to allow this special expansion of concepts relevant to plasma techniques. It is resumed at BOX B8Y.
	. * Add to BOX B8 letters A/X following BRV; eg
B8B	. Flow of plasmas
	. . Rotation
B8D N	. . . Rotating plasmas

[Particles BM]
[Nuclei BO]
[Energy interactions & forms BOAF]
. . . [Nuclear fusion BOXB]
. . . . [Special techniques]
. . . . . [Plasma reactions]
. . . . [Rotating plasmas BOXB8DN]

	. . . . . Plasma systems
BOX B8V L	. . . . . . Collisionless plasmas
B8Y	. . . . *Other investigative methods*
	. . . . . * Normal retroactive synthesis is resumed here after its interruption at BOX B8.
	. . . . . * Add to BOX B8Y numbers 7W/9 following B if applicable.
	. . . . *Nuclear reactions, types of fusion*
	. . . . . *By product nucleus*
BSH J	. . . . . . Helium (fusion product)
BSH L	. . . . . . Lithium (fusion product)
	. . . . . *By projectile*
BTH B	. . . . . . Hydrogen fusion projectiles
BTH C	. . . . Hydrogen-1 (fusion projectile)
BTH D	. . . . Deuteron, hydrogen-2 (fusion projectile)
BTH E	. . . . Triton, hydrogen-3 (fusion projectile)
	Types of nuclei
	. * See second note under BO.
	. * For Nuclides, see BPV
D	. *By other particles*
	. . * Add (retroactively) to BOX D letters A/X following BN if applicable.
	. . * Add (retroactively) to BOX E letters O/Q following B if applicable.
DDR H	. . Stable nuclei
DDR J	. . Unstable, radioactive nuclei
	. . . * See also compound nuclear processes BOR X
DDR K	. . . Daughter products
DDR L	. . . Fissile nuclei
DDR M	. . . Fissionable nuclei
DDR N	. . . Radioactive series, nuclear series, nuclear decay series, radioactive decay series
DLD	. . Low energy nuclei
	. *By various characteristics*
FP	. . Isotopes (nuclei)
FQ	. . Active nuclei
FT	. . . Hypernuclei
	. . . . * Extremely unstable particles produced when a lambda particle replaces a neutron in a nucleus.
	. *By mass number*
GA	. . Mass number nuclei
GB	. . Nuclei with mass numbers 1-5
GC	. . . Few-nucleon systems
GD	. . Nuclei with mass numbers 6-19
GG	. . Nuclei with mass numbers 20-38

Types of nuclei

[Matter BLY]
[Particles BM]
[Atomic & nuclear physics together BNY]
. [Nuclei BO]
. . [Types of nuclei]
. . . [By mass number]
. . . . [Nuclei with mass numbers 20-38 BOXGG]

BOX GJ . . . . Nuclei with mass numbers 39-58
GL . . . . Nuclei with mass numbers 59-89
GN . . . . Nuclei with mass numbers 90-149
GQ . . . . Nuclei with mass numbers 150-189
GS . . . . Nuclei with mass numbers 190-219
GV . . . . Nuclei with mass numbers 220 or more
GW . . . . . Superheavy nuclei
. . . *By element*
HB . . . . Hydrogen nucleus
HD . . . . . Deuterons, deutons
. . . . . . * For deuterons as product nuclei, see BOS HD; as targets, see BOV HD.
. . . . . . *Interactions with other particles*
HDQ YO . . . . Deuteron-nucleus interaction
HDR DC . . . . . . Cosmic ray deuterons
HE . . . . . Tritons
. . . . . . *Interactions with other particles*
HEQ YO . . . . Triton-nucleus interaction
. . . . . . Hypernuclei
. . . . * A lambda particle can replace the neutron in a nucleus to form a very unstable hypernucleus.
HEX FT . . . . Hypertriton
HH . . . . Helium nucleus, alpha-particles
HHR DC . . . . . Cosmic ray alpha particles
HI . . . . . Helium-3 nucleus
. . . . . . *Interactions with other particles*
HIQ YO . . . . Helium-3-nucleus interactions
HJ . . . . . Helium-4 nucleus
HL . . . . Lithium nucleus
. . . . *Other nuclei, by element*
. . . . . * Add to BOX letters I/Q following C Chemistry, except for thorium, uranium & plutonium, which are notated separately; eg
IU . . . . . Beryllium nucleus
JC . . . . . Magnesium nucleus
JK . . . . . Carbon nucleus
KL . . . . . Chlorine nucleus
P . . . . . Thorium nucleus
Q . . . . . Uranium nucleus
R . . . . . Plutonium nucleus
BOY Atomic & molecular & ion physics (together)
. * Add to BOY letters A/Y following BP so far as applicable; eg
BOY BD . Energy levels
NV . Basic interactions
PL . . Weak interactions

[Physics B]
[Matter BLY]
[Particles BM]
[Atomic & molecular & ion physics together BOY]
. [Basic interactions BOYNV]
. . [Weak interactions BOYPL]

BOY PLB G . . . Van der Waals forces

BP Atoms (physics), atomic physics, single atoms (physicsa of)
. * Structure, processes and properties of the atom regarded as a whole. For the atom as a constituent of molecular reactions, see Chemistry (Class C).
. * Add to BP as instructed in the notes under BNB.
BP8 MF . Quantum field theory
BP9 D . Size
DK . . Atomic radii
BPB B . Energy
. . * For atomic energy, see Nuclear energy BOB B.
D . . Energy levels (atoms), electron energy states (atoms)
. . . * Narrowly, the energy state of a particular electron in orbit around the central nucleus of an atom. More broadly, the possible energy value of an electron or nuclear particle.
. . . * Strictly, the state can be considered separately from the level, although they are usually treated as synonymous. If state is distinguished from level, use BPB E Quantum states.
. . . * For energy levels of particular particles, see the particle; eg Molecules - Charge transfer state BQB EW; for energy bands, see condensed matter BTX BF.
. . . *Processes & properties*
. . . . * For degeneracy, see degenerate level BPB EQ.
. . . . * Add to BPB D numbers & letters 2/G following B; eg
D9D J . . . . Energy level width
DFP . . . . Polarization
DJ . . . . Electron density of state
DK . . . . Fermi level, Fermi energy
. . . . . * Probability of locating electrons in a level.
DL . . . . Electron correlations
DM . . . . Population inversion
. . . . . * When higher energy state has more electrons than a lower state.
. . . . . * For pumping, see Lasers (techniques) B6K SFI O.
DN . . . . Transitions
. . . . . * See also radioactivity BOF K

Atoms (physics)

[Matter BLY]
[Particles BM]
[Atoms physics BP]
[Energy BPBB]
. [Energy levels atoms BPBD]
. . [Processes & properties]
. . . [Transitions BPBDN]

BPB DN8 MF . . . . Sum rules, quantized electron transition
DQ . . . Isomerism
. . *Structure*
DT . . . Orbitals, electron configuration (atoms)
DT8 MT . . . . Principal quantum number
DU . . . . Electron shells (atoms)
DUL . . . . Larmor orbit, Larmor precession
. . *Types of energy states*
E . . . Stationary state
EG . . . Ground state
. . . . * See also atomic orbitals, BPB DT
EK . . . Excited state
EKL . . . . Doubly excited states (atoms)
EKM . . . . . Auger effect, Auger ionization
. . . . . . * Loss of energy on ionization.
EKN . . . . Atomic metastable state
EKR . . . . Atomic resonant state
EKS . . . . . Flashback resonance
EKT . . . . Metastable state
EM . . . Bound state
. . . . Mathematics
EM2 M9N . . . . . Bethe-Salpeter equations
EP . . . Unbound state
EQ . . . Degenerate level
EQP . . . . Degeneracy
ER . . . Fine state, atomic fine structure
ES . . . Hyperfine state
ET . . . . Isotope shift
. . *Types of energy levels special to a particle, etc.*
EU . . . Multiplets
EW . . . . Atomic triplet state
. . . . . * Two unpaired electrons.
F . . Energy bands
. . . * See condensed matter BTX BF
. Force
G . . Atomic force, interatomic potential
BPC H Statics
J . Atomic mass, atomic weight
. . * For mass as quantum number, see BPM H; for isotopes, see BPW.
JM . . Relative abundance
BPD N Rotation
BPE Y Radiation
. Spectra

[Physics B]
[Matter BLY]
[Particles BM]
[Atoms physics BP]
[Radiation BPEY]
. [Spectra]

BPF E . . Atomic spectra
. . . * For atomic energy level transitions, see BPB DN; for hyperfine structure, see BPB ES.
E9D J . . . Atomic spectral line breadth
HL . Atomic fluorescence
IN . Excitation
. . * For excited state (energy levels), see BPB EK.
IXP . . Electron impact excitation (atoms)
P . Polarization
S . Collision
. . * For collisions between two specified particles, see particle interractions BPQ.
SM . . Atomic inelastic collisions
BPG B . Atomic beams
BPH IBH Electric field
IBH FP . Polarizability
K Atomic charge
KO . Electric moment
. . * See BPM MP
BPJ BH Magnetic field
JC Magnetic moment
BPK J Ionizing radiation
JFC . Ionization
Quantum properties
BPM MP . Electric moment
BPN V Basic interactions
BPP L . Weak interactions
LBG . . Van der Waals forces
Interactions with other particles
. * For molecule-atom interactions, see BQQ YP; for ion-atom interaction, see BQU QYP.
. * Add to BPQ letters B/Y following BN;
. * Add to BPQ Y letters O/Q following B; eg
BPQ P . Electron-atom interaction
PFI N . . Excitation
PFS . . Collision
PFT . . Scattering
PRF . Positron-atom interaction
PRF FT . . Positron-atom scattering
YP . Atom-atom interaction
YV *Parts of atoms*
. * For electrons, see BNP; for nucleons, see BNU; for nucleus, see BO.

[Physics B]
[Matter BLY]
[Particles BM]
[Atoms physics BP]
[Parts of atoms BPQYV]

Types of atoms
BPR B . *By properties other than quantum properties*
BR . . Exotic atoms
. . . * Unstable atoms in which electrons are replaced by another negative particle.
. . . * See also Positronium BNP RFT
. *By quantum property*
CB . . Symmetry groups (atoms)
. *By aggregation*
DRC . . Atomic clusters
. *By charge*
H . . Charged atoms
. . . * See Ions BQU
HW . . Neutral atoms, uncharged atoms
. *By mass*
LR . . Medium-heavy atoms
LS . . Heavy atoms
. *By other particles*
N . . Muonic atoms
Q . . Hadronic atoms
S . . . Mesic atoms
. *By individual element*
. . * Add to BPU letters I/Q following C Chemistry (but notation is provisional); eg
BPU HB . . Hydrogen atom
HBR HW . . . Hydrogen neutral atom
HH . . Helium atom
BPV . Nuclides, nuclear species
. . * Atoms characterized by their atomic number, mass number and the constitution of their nucleus.
. . Graphs
BPV 2HU . . . Segre charts
. . *Types*
R . . . Mirror nuclides
S . . . Wigner nuclides
BPW . . . Isotopes, isotopic nuclides
. . . . * Two or more nuclides which have identical nuclear charge (atomic number) but differ in nuclear mass.
. . . . *Operations*
BPW 73P . . . . . Isotope separation
73Q . . . . . . Electrolytic separation (isotope)
73R . . . . . . Gas diffusion separation (isotopes)
73S . . . . . . Centrifugation (isotopes)
BPX . . . . Radioisotopes, radioactive isotopes, radioactive nuclides, radionuclides
. . . . . * For nuclear decay series, see BOX FS.
BPX R . . . . . Stable isotopes

[Particles BM]
[Atoms physics BP]
. [Types of atoms]
. . [Nuclides BPV]
. . . [Types]
. . . . [Isotopes BPW]
. . . . . . [Stable isotopes BPXR]

BPX S . . . . Fissionable nuclides
T . . . . Fissile nuclides
BPY . . . Individual nuclides
. . . . * Specified by mass number as follows:
. . . . * Add to BPY letters A/W following BOX G; eg BPY A Nuclides with mass numbers 1-5.

BQ Molecules, molecular physics, single molecules
. * See also Stereochemistry C
. * Add to BQ as instructed in the notes under BNB.
BQ9 D . Size
DS . . Molecular volume
JV . Configuration
. . * Use BQB DV
. Energy levels, energy states
BQB D . . Electronic structure
. . . * For valency, see Chemistry C
DQ . . . Isomerism
. . . . * Having the same atomic and mass numbers but with different energy states.
. . . . * For stereoisomerism, see Stereochemistry C
DQM . . . . Molecular rotational isomerism, molecular internal rotation
DT . . . Molecular orbitals, wave function (molecules
. . . . Calculation
DT7 5 . . . . . Molecular orbital calculation
DT7 5A . . . . . . Specific calculations
. . . . * Arrange A/Z; eg NDO calculation BQB DT7 5N.
DV . . . . Molecular configurations
DW . . . Molecular orientation, molecular alignment
. . . *Types of energy states*
EKT . . . . Molecular metastable state
ER . . . Fine structure
ES . . . Hyperfine structure
. . . *Special types of energy level*
EX . . . . Charge transfer state
IM . . Molecular moment
. . Internal forces
BQC D . . . Bond energy
F . . Energy loss
. . . * See also Spin BQM K
FH . . . Stopping power
. . . . * Energy loss per molecule normal to the motion of the particle travelling a distance.
H . Statics
. . Mass

[Physics B]
[Matter BLY]
[Particles BM]
[Molecules BQ]
[Statics BQCH]
. [Mass]

BQC J	. . Molecular weight
BQD N	Rotation
	. * For internal rotation, see BQB DQM.
NX	. Libration
V	Vibration
	. Energy states
VBD	. . Molecular vibronic states
BQF E	Spectra
	. Breadth
E9D J	. . Molecular spectral line breadth
IN	Excitation
	. Dexcitation
IQ	. . Internal conversion (molecules)
IR	. . . Non-radiative transition
IXP	. Electron impact excitation
	Decay
J	. Dissociation (molecules)
	. . * Breakdown of molecules into smaller
	. . molecules or atoms.
M	Refraction
O	Resonance
	Polarization
P	. Molecular polarizability
S	Collision
SM	. Molecular inelastic collision
SR	. Energy transfer collision (molecular)
T	Scattering
TJL	. Molecular elastic scattering
	Beams
BQG B	. Molecular beams
BQH K	Charge
	Magnetism(
BQJ KS	. Molecular magnetic susceptibility
BQM	Quantum number properties
BQM K	. Spin
BQN V	Basic interactions
BQP L	. Weak interactions
LBG	. . Van der Waals forces (molecular)
	Interactions with other particles
	. * For chemical reactions see Chemistry (Class C)
	. * Add to BQQ letters B/X following BN;
	. * Add to BQQY letters O/Q following B; eg
BQQ W	. Neutron-molecule interaction
WFR U	. . Coupling
WFT	. . Scattering
YP	. Atom-molecule interaction
YQ	. Molecule-molecule interaction
	. Ion-molecule interactions
	. . * See Ion physics BQU QYQ
	Types of molecules
BQT D	. Diatomic molecules

[Physics B]
[Matter BLY]
[Particles BM]
[Molecules BQ]
. [Types of molecules]
. . [Diatomic molecules BQTD]

BQT E	. . Polyatomic molecules
G	. . Conjugated molecules
M	. . Macromolecules
	. . . Energy states
MBD	. . . . Macromolecular energy state
BQU	Ions, ion physics
	. * Electrically charged atoms, molecules or groups of
	. atoms or molecules.
	. * Add to BQU as instructed in the notes under BNB.
BQU AG	. Thermodynamics
	. . * See also Ion temperature BQU GV
BD	. Energy levels
	. . * For ionized state, see BKJ FC.
CH	. Statics
CJ	. . Mass
CJ9 2H	. . . Change of mass ions
	. . Density
CLM	. . . Ion concentration
	. Velocity
DC	. . Ion mobility
DU	. Oscillation
	. . * See also ion optics BQU F
	. Radiation, wave motion
	. . * For ionization, see BKJ
F	. . Ion optics
FG	. . . Emission
FGT	. . . . Secondary ionic emission
FGU	. . . . Thermionic ionic emission
FIN	. . . Excitation
	. . . . * See also Metastable ions BQU TIU
	. . . Decay
FJ	. . . . Dissociation (ions)
FS	. . . Collision
	. . . . * See also Recoil ions BQU TS
FT	. . . Scattering
FUF	. . . . Diffusion
	. . . . Beams
GB	. . . . . Ion beams
GV	. Temperature
HIB H	. Electric fields
HK	. Charge
HL	. Potential
	. Current
HP	. . Ionization current
L	. Ion optics
	. . * Use BQU F.
Q	. Interactions with other particles
	. . * Add to BQU Q letters B/Y following BN;
	. . * Add to BQU QY letters O/Q following B: eg

[Physics B]
[Matter BLY]
. [Particles BM]
. . [Ions BQU]
. . . [Interactions with other particles BQUQ]

BQU QYP . . . . Atom-ion interaction
QYQ . . . . Molecule-ion interaction
QYQ FS . . . . . Molecule-ion collisions
. . . *Types of ions*
. . . . *By charge*
RHU . . . . . Cations, positive ions
RHV . . . . . Anions,negative ions
RHW . . . . . Ion pairs, paired ions
RHY . . . . . Stripped atoms
. . . . . . * Ionized atoms from which at least one
. . . . . . electron has been removed.
RI . . . . Magnetic ions
RJ . . . . Complex ions
S . . . . Heavy ions
. . . . . *Interactions with other particles*
SQY O . . . . . . Heavy ion-nucleus interaction
T . . . . *By origin*
. . . . . * Add to BQU T letters A/Y following BF
. . . . . * Add to BUQ U letters A/F following
. . . . . BG; eg
. . . . . Excitation
TIU . . . . . . Metastable ions
. . . . . Collision
TS . . . . . . Recoil ions
UH . . . . . Electrolytic ions
BQV F . . . . Free ions
BQW . . . . *By element*
. . . . . * Add to BQW letters I/Q following C
. . . . . Chemistry; eg
BQW HA . . . . . Metal ions
HB . . . . . Hydrogen ion
BQX Vacuum physics
. * See also Vacuum techniques B6Q X

[Physics B]
[Vacuum physics BQX]

BR Bulk matter physics, substances (bulk matter), macrophysics
. * Physics of aggregates of molecules (including
. plasmas).

BR3 6 . Practical physics
B . . Equipment & materials
BR4 . . . Instrumentation
BR6 2 . . Investigative techniques
9 . . Physical methods
BR7 6 . . Measurement
BR8 B . Theoretical physics
. *General properties*
BR9 D . . Spatial dimensions
DEK . . Isotropy
DEL . . Anisotropy
G . . Systems characteristics
J . . . Structure
JV . . . . Shape, configuration
Q . . . Control systems
QG . . . . Open loop systems
QJ . . . . Closed loop systems, feedback control
. . . . systems
BRA F . Energy interactions & forms
. . * Add to BR letters A/L following B amplified as
. . indicated below.
G . . Thermodynamics
. . . * In the sense, somewhat narrower than that at
. . . BAG, of the transfer of energy to, from and
. . . between macroscopic bodies.
. . . * For thermal properties of bulk matter, see
. . . BRG P.
T . . . Transport processes & properties
. . . . * Most of the literature refers to these
. . . . properties in particular energy forms or states
. . . . of matter, q.v. This class takes only general
. . . . works on transport properties in bulk matter.
. . . . * For mass transfer, see BAX.
T2M 9L . . . . Equations
T2N 9N . . . . . Bolzmann equation, transport
. . . . . equation
V . . . . Diffusion general
X . . . . Mass transfer
. . . . . * See also Heat transfer BRG Q
BRB . . Mechanics
BRB G . . . Forces
IM . . . . Moments
J . . . . Pressure
JBH . . . . . Pressure field
JP . . . . . Low pressure physics
JQ . . . . . High pressure physics
JW . . . . External forces

Periodic motion

[Physics B]
[Bulk matter physics BR]
[Energy interactions & forms BRAF]
[Mechanics BRB]
[Forces BRBG]
. [External forces BRBJW]

BRB K . . Deformation
. . . * For works on deformation in bulk matter
. . . in general; most of the literature concerns
. . . the solid state and the detailed schedule is
. . . given at BVB K.
. . . * Add to BR letters BK/CB following BV; eg
KL . . . Stress/strain relations
L . . . . Stress
P . . . . . Shear
T . . . Strain
BRC B . . . Elasticity
CC . . . . Viscoelasticity
H Statics
L . Density
N . Equilibrium
P . Stability
S Motion
X Dynamics
BRD A . Kinematics
C . . Velocity
D . . Acceleration
E . Kinetics
. . * For special elements & attributes of flow, see
. . Fluids BSG E.
. . Theory
E8E K . . . Kinetic theory
S . Periodic motion
TB . . Harmonics
BRE . . Mechanical vibration & oscillation,
. . sonics
. . . * The term Sonics is sometimes used for
. . . sound waves narrowly, in which case use
. . . BRG H.
. . . * Add to BRE A letters A/DS following B;
. . . * Add to BRE letters B/Y following BE; eg
. . . *Operations*
BRE 72 . . . . Control
72B . . . . . Hysteresis
72D . . . . . Damping
72E . . . . . . Isolation damping
73D . . . . Generation, reproduction
. . . . . * As an operation in investigation.
. . . . . * For synthesis of sound, see
. . . . . BRG H73 N
. . . *Properties & processes*
B . . . . Harmonics
BH . . . . . Nodes
BQ . . . . . Anharmonics
D . . . . Frequency
DD . . . . . Mode
. . . *Types of vibrations*

[Bulk matter physics BR]
[Energy interactions & forms BRAF]
[Mechanics BRB]
[Dynamics BRCX]
[Periodic motion BRDS]
[Mechanical vibration & oscillation BRE]
. [Types of vibrations]

. . * For vibration in a particular state of
. . matter, see latter; eg BTE Vibration &
. . oscillation in gases.
BRE VV . . *By property*
. . *By direction/transience &*
. . *waveform*
WJ . . . Linear vibration, non-sinusoidal
. . . vibration
WL . . . Non-linear vibration, sinusoidal
. . . vibration, angular
. . . vibration, circular vibration
WM . . . . Random vibration
WN . . . . . Chatter
WP . . . Rotational vibration
WPQ . . . . Critical speed
WPR . . . . Hunting
WR . . . Transverse vibration
WS . . . . Shear vibration
. . . . . * See also S waves (seismology) DG
WU . . . Longitudinal vibration
WV . . . . Axial vibration
XC . . *By degree of freedom*
. . *By internal/external origin*
XG . . . Free vibration, natural frequency
XH . . . Forced vibration
XL . . *By origin as to action*
. . . * Arrange A/Z (eg blowing, plucking).
. . *By origin as to body vibrating*
XM . . . Vibrating bodies
. . . . * For acoustic phenomena in
. . . . particular types of materials, see
. . . . BRG J.
. . . . *By solid medium*
XN . . . . . Linear bodies (vibration)
XNG . . . . . . Strings
XNH . . . . . . Rods
XNJ . . . . Straight rods
XNK . . . . Curved rods
XNN . . . . . . Tuning forks
XP . . . . . Two-dimensional bodies
. . . . . (vibration)
XPM . . . . . . Membranes (vibration)
XPN . . . . . . Diaphragms (vibration)
XPP . . . . . . Plates (vibration)
XS . . . . . Three-dimensional bodies
. . . . . (vibration)
XSP . . . . . . Pendulums
XT . . . . *By specific medium*
. . . . . * Arrange A/Z (eg bells, foghorns).
BRF Waves

Acoustics

[Physics B]
[Bulk matter physics BR]
[Energy interactions & forms BRAF]
[Mechanics BRB]
[Dynamics BRCX]
[Periodic motion BRDS]
[Waves BRF]

. * Division of waves is parallel with that of vibrations and oscillations (see notes at BE and BF). For bulk matter, some amplification may be made here by adding letters following BRE.
. * Add to BRF letters A/Y following BRE so far as is necessary; eg

	. *Properties*
BRF D	. . Frequency
	. *Types*
YG	. . Standing waves
YR	. . Transverse waves
YS	. . . Shear waves
	. . . . * See also S waves (seismology) DG
YU	. . Longitudinal wave
BRG H	Acoustics, sound waves, sonics (sound)
	. * For acoustic phenomena in a particular medium (gas, liquid, solid, etc.) see medium.
H36	. *Practical & experimental*
H72 N	. . Analysis of sound
H72 P	. . . Qualitative analysis of sound
H72 Q	. . . Quantitative analysis of sound
H73	. . Generation of sound
	. . . * See also Types of sound by source BRG I
H73 D	. . Reproduction of sound
	. . . * Purely acoustic studies; most of the literature will go under sound recording & reproduction in Technology Class U.
H73 L	. . . *By causative action*
	. . . . * Arrange A/Z; eg
H73 LB	. . . . Blowing
H73 M	. . . . *By specific body*
	. . . . . * Arrange A/Z (eg bells, foghorns).
H73 N	. . . Synthesis of sound
H74 J	. . Sensing & detection
H7M	. . Acoustic spectroscopy
H8B	. Theory
	. *Properties & processes*
H9D E	. . Directionality
HBJ	. . Pressure
HDT	. . Velocity

[Energy interactions & forms BRAF]
[Mechanics BRB]
[Dynamics BRCX]
[Periodic motion BRDS]
[Acoustics BRGH]
[Properties & processes]
. [Velocity BRGHDT]

BRG HFB	. Harmonics
HFB K	. . Quality of sound, timbre
HFC	. Propagation, transmission
HFD	. Frequency
	. *Special acoustic attributes*
HFE N	. . Pitch
HFE O	. . . Octaves
HFE P	. . Tone
HFE Q	. . . Beats
	. . . . * Subjective difference in tone associated with two closely similar frequencies being heard together.
	. . . . * See also Interference BRG HFR
	. Amplitude
HFE S	. . Loudness, volume (sound), subjective intensity (sound)
HFN	. Reflection
HFN R	. . Reverberation (acoustics)
HFN S	. . . Albedo (acoustics)
HFO	. Resonance
	. . * For magnetic resonance, see BRG HJF O.
HFO S	. . Dissonance, consonance
	. . . * See also Noise BRG HN
HFR	. Interference
HFX X	. Pulses
HFY G	. Standing waves
	. *Interactions with other energy forms*
	. . * For Acoustooptics, see BRL GH
HGP	. . Thermal acoustics
HH	. . Electroacoustics
	. . . * Effect of electrical phenomena on sound.
	. . . * See also Acoustoelectric effects BVI UGH
HJ	. . Magnetism
HJF O	. . . Magnetic resonance (acoustics)
HL	. . Acoustooptics, acoustic optics
HM	. *Relations with particle physics*
	. . * Add to BRG HM letters M/Q if applicable.
	Types of sound
HN	. Noise (acoustics)
	. . * Subjective category of sounds which are undesired by the (human) recipient.
HP	. . Reduction of noise
HQ	. . . Insulation of sound
	. . . . * Most of the literature goes in technology.

[Dynamics BRCX]
[Periodic motion BRDS]
[Acoustics BRGH]
[Types of sound]
[Noise acoustics BRGHN]
. [Reduction of noise BRGHP]
. . [Insulation of sound BRGHQ]

BRG HR . Ambient noise
HS . Cavitation noise
. . * For sonic boom, see BTM GHT.
HW . White noise
. . * Aperiodic sound.
I *By source*
. * Alternative (not recommended) for libraries wishing to keep together all acoustic studies.
. * Add to BRG I numbers & letters 3/9,A/Z from the whole classification; eg physics of music BRG IWV.
J *By medium*
. * Alternative (not recommended) for libraries wishing to keep together acoustic studies in all media (see note at head of this class).
. * Add to BRG J letters R/W following B; eg acoustic properties of solids BRG JV.
By frequency
K . Acoustic spectrum
K9L R . . Linear
K9M . . Non-linear
L . Infrasound
M . Audible sound
. . * Usually assumed; see note at BRG HN.
N . High-frequency sound, ultrasonics
O . . Extremely high frequency sound
. . . * For phonons, see solid state BVG O8N.

[Bulk matter physics BR]
[Energy interactions & forms BRAF]
[Mechanics BRB]
. . . . [Types of sound]
. . . . . [By frequency]
. . . . . . [High-frequency sound BRGN]
. . . . [Extremely high frequency sound BRGO]

BRG P Thermal properties of bulk matter
P93 . *Environmental conditions*
P94 F . . Constant volume, isochoric, isometric
P94 K . . Constant pressure, isobaric
P94 Q . . Constant temperature, isothermal
PAG . Thermodynamics
. . *Processes & properties*
PAP D . . . Reversible
PAP E . . . Irreversible
PAP Q . . . Adiabatic
PAP T . . . Isothermal
PAP V . . . Isentropic
PAP Y . . . Cycles
PAQ . . . . Named cycles
. . . . . * Arrange A/Z; eg
PAQ C . . . . . Carnot cycle
PAQ R . . . . . Rankine cycle
PAT . . . Transport processes
. . . . * For heat transfer, see BRG Q.
PBB . Energy
PCN . Equilibrium
PCN P . . Balancing
PDE . Kinetic energy
. . * For temperature, see BRG V; for combustion, see Chemistry C.
PM . *Relations with particle physics*
. . * Add to BRG PM letters M/Q following B if applicable.
. *Special thermal properties & processes*
PQ . . Heat, quantity of heat
. . . * For latent heat, see Phase transormation BRN S.
. . . Measurement
PQ7 6 . . . . Calorimetry
PQ7 64 . . . . . Calorimeters
PQ7 7 . . . . . Calorimetric units
PQ7 8R . . . . . Bomb calorimetry
PQ7 8S . . . . . Continuous flow calorimetry
PQ7 8V . . . . . Microcalorimetry
PR . . . Heat capacity, thermal capacity
PS . . . . Specific heat capacity
Q . . Heat transfer, heat exchange
. . . * Includes heat & mass transfer together; for mass transfer alone, see transport processes BRA X;
. . . * See also Chemical engineering V
. . . * Add to BRG QA letters A/J following B.
QK . . . Radiation

Thermal properties of bulk matter

[Physics B]
[Bulk matter physics BR]
[Energy interactions & forms BRAF]
[Thermal properties of bulk matter BRGP]
[Special thermal properties & processes]
[Heat transfer BRGQ]
[Radiation BRGQK]

. * See BRG R
BRG QL Heat flux, transmission (heat flux), heat flow
QL2 QK . Heat transfer coefficient
QL7 6 . Measurement
QL7 64 . . Heat flow meters
QLM . Direct flow (heat)
QLP . Counter flow (heat)
QM Thermal cycling
By process
QN . Conduction of heat
. . Coefficient of thermal conduction
QN2 QK . . . Thermal conductance
QNT . . Thermal conductivity
QP . . Diffusion of heat
QR . Heat resistance, resistivity
QT . Thermal transpiration, thermal effusion
. . * See Gases BTM GQT
QV . Convection
. . * Heat transfer in a fluid by the movement of the fluid itself; see Fluids BSM GQV.
R . Thermal radiation, heat radiation
. . * Add to BRG R letters K/L following B so far as applicable.
RS . . Black body radiation
RT Thermal emission
RTS . Emissivity
RU Thermal absorption
. * For black body radiation, see BRG RS.
By resulting temperature
T . Heat loss, cooling
. . * For supercooling, see change of state BRN P94 T.
. . Radiation
TEX . . . Radiation cooling
TSH . . Thermoelectric cooling
TSJ . . Magnetic cooling
TT . . Phase change cooling
TTS . . . Linde cooling
TTV . . Film cooling
TTX . . Expansion cooling
TTY . . Sweat cooling
TU . . Evaporative cooling
TV . . . Tranpiration cooling
TVS . . . Boiling heat transfer

[Thermal properties of bulk matter BRGP]
[Special thermal properties & processes]
[Heat transfer BRGQ]
. [By resulting temperature]
. . [Heat loss BRGT]
. . . [Evaporative cooling BRGTU]
. . . . [Boiling heat transfer BRGTVS]

BRG TW . . . Ablation cooling
TX . . . Sublimation cooling
TY . . . Other forms of cooling
. . . . * Arrange A/Z.
U . . Heat gain, heating
. . . * For superheating, see change of state BRN P94W.
UEY . . . Radiation
UQN . . . Conduction
V Temperature
. * Property of an object which determines the direction of heat flow when in thermal contact with another body.
. * Add to BRG VA letters A/Q folowing B if applicable.
V72 . Control
V72 4 . . Thermostats
V72 4AC . . . Electrical thermostats
V72 4GA . . . Thermistors
V72 4V . . . Gas thermostats
V72 4W . . . Cryostats
V72 4X . . Pneumatic temperature control
V72 4Y . . Other control methods
. Measurement
V76 . . Thermometry
V76 4 . . . Thermometers
V76 47A . . . . Testing & correcting
V76 47A H . . . . . Hypsometers
V76 5C . . . . Electric thermometers
V76 5D . . . . . Resistance electric thermometers
V76 5E . . . . . . Special types
. . . . * Arrange A/Z; eg thermistor.
V76 5G . . . . . Thermocouples (thermometers)
V76 5H . . . . . . Thermopiles
V76 5J . . . . Gas thermometers
V76 5K . . . . Vapour pressure thermometers
V76 5M . . . . Mercury thermometers
V76 5N . . . . . Beckmann thermometer
V76 5P . . . . Other liquid expansion thermometers
V76 5R . . . . Solid expansion thermometers
V76 5S . . . . . Bimetallic thermometers
V76 5T . . . . . Other solid expansion thermometers
. . . . . . * Arrange A/Z; eg quartz thermometers BRG U76 5TQ
V78 L . . Temperature scales
V78 LT . . . Thermodynamic temperature, absolute temperature scales
V78 M . . . . Kelvin scale

[Bulk matter physics BR]
[Energy interactions & forms BRAF]
[Thermal properties of bulk matter BRGP]
. [Special thermal properties & processes]
. . [Temperature BRGV]
. . . [Measurement]
. . . . [Temperature scales BRGV78L]
. . . . . [Thermodynamic temperature BRGV78LT]
. . . . . . [Kelvin scale BRGV78M]

BRG V78 MR	. . . . . Relative temperature scales, practical temperature scales
V78 N	. . . . . . Celsius scale, Centigrade scale
V78 O	. . . . . . Fahrenheit scale
V78 P	. . . . . . Reaumur scale
V7M	. . . . . Spectral methods (thermometry)
V7N SL	. . . . . . Colour thermometry, thermocolour methods
V7T	. . . . . Acoustic methods (thermometry)
V7V	. . . . . Calorimetric thermometry
V92 H	. . . . Temperature variations
VJ9 2J	. . . . . Temperature gradients
VJ9 4R	. . . . . Adiabatic conditions
	. . . *Types of temperature*
VN	. . . . Inversion temperature
VR	. . . . Normal temperature, room temperature, ambient temperature
VS	. . . . Critical temperature
VSR	. . . . . Minimum temperature
VSX	. . . . . Maximum temperature
VW	. . . . Low temperatures (general)
	. . . . . * For cryogenics, see
VX	. . . . High temperatures (general)
	. . . . . * For high temperature physics, see BRG X.
	. *Thermal regimes*
W	. . Low temperature physics, cryogenics
	. . . Measurement
W76	. . . . Cryometry
WX	. . . Absolute zero temperature, zero temperature
X	. . High temperature physics (general)
	. . . Measurement
X76	. . . . Pyrometry
X76 T	. . . . . Resistance pyrometry
X76 U	. . . . . Radiation pyrometry, total radiation pyrometry
X76 V	. . . . . . Optical pyrometry
X76 W	. . . . Heat radiation pyrometry
X76 X	. . . . Light radiation pyrometry
Y	Electrical & magnetic properties
	. * The detailed schedule for this class is given under solid state BVG Y.
	. * Add to BR letters H/K following BV; eg
BRH U	. Conducting systems (electrical)
BRJ	. Magnetic properties

[Physics B]
[Bulk matter physics BR]
[Energy interactions & forms BRAF]
[Electrical & magnetic properties BRGY]
[Magnetic properties BRJ]

BRJ FO	. Magnetic resonance
BRK	Electromagnetic radiation
	E-m waves by frequency & wavelength
BRK M	. Radiofrequency waves
U	. Microwaves
BRL	. Optical properties of bulk matter, optics (bulk matter)
	. . * Most of the literature refers to visible light (BRL V), details for which are taken from the general class for optical properties of bulk matter (BRL 2/BRL Q).
	. . * An alternative (not recommended) is to confound visible light with optics in general and use BRL 2/BRL Q for optics in general and visible light.
BRL 36	. . Practical & experimental optics
3B	. . . Materials & equipment
	. . . . Undesired effects
3B3 7I	. . . . . Aberration (optics)
	. . . . *Operations on materials & equipment*
3BK	. . . . . Testing & monitoring
	. . . . *Properties*
	. . . . . *Special to optical materials*
3ML	. . . . . . Reflectivity (equipment properties)
3R	. . . . Materials
3RQ V	. . . . . Optical film
3RQ X	. . . . . Coatings
3TF	. . . . . Optical fibres
	. . . . . . * For fibre optics, see B6L FCT.
3TG	. . . . . Optical glass
3TP	. . . . . Light sensitive materials
3U	. . . . Equipment & plant
	. . . . . * Equipment serving particular processes, etc. goes with the process. But although some of the devices below primarily perform a particular operation in optics, (eg mirrors reflect) they may appear in many different contexts and are therefore treated as general accesories.
	. . . . . * When used in a particular process, wave-length, etc. see latter; eg Absorption - Filters BRL FL4 UF. Filters BK
	. . . . . * When treated as a component in a containing instrument, see latter; eg microscope mirrors BRL 7J4 WM.
	. . . . . * For light sources, see Production BRL 73; see also Physical optics BRL F

Optical properties of bulk matter

[Energy interactions & forms BRAF]
[Electrical & magnetic properties BRGY]
[E-m waves by frequency & wavelength]
[Optical properties of bulk matter BRL]
[Practical & experimental optics BRL36]
[Materials & equipment BRL3B]
[Equipment & plant BRL3U]

BRL 4	Optical instruments
45	. Optical components
	. . * Use BRL 4U/X.
	. . *Special to optics*
4UF	. . . Optical filters, light filters
	. . . . * For filters controlling particular processes, etc, see latter - eg Absorption - Filters BRL FL4 UF.
4UF P	. . . . Spatial filters
4UF S	. . . . Spectral filters
4UH	. . . Luminescence devices
	. . . . * See also Phosphors BRL FH3 TP
4UH L	. . . . Fluorescent screens
4UR	. . . Rangefinders (optical instruments)
4UY	. . . Lenses & prisms (together)
4V	. . . . Lenses
4V3 7I	. . . . . Aberrations
4V3 7IK	. . . . . . Image error (lenses)
4V3 7IL	. . . . Aperture error
4V3 7IM	. . . . Sperical aberration
4V3 7IN	. . . . Coma (lenses)
4V3 7IP	. . . . Curvature of field
4V3 7IQ	. . . . Astigmatism (lenses)
4V3 7IR	. . . . Distortion (lenses)
4V3 7IS	. . . . Chromatic aberration, chromatism
	. . . . . *Properties*
4V3 JM	. . . . . . Principal focus
4V3 JN	. . . . . . Principal focal plane
4V3 JP	. . . . . . Focal length, refractivity of lens system
4V3 JR	. . . . . . Field of view, visual field, image field
	. . . . . *Parts*
4V3 NB	. . . . . . Surfaces (lenses)
4V3 NC	. . . . Bloomed surfaces, coated surfaces (lenses), blooming

[Optical properties of bulk matter BRL]
[Special to optics]
[Lenses & prisms together BRL4UY]
[Lenses BRL4V]
. [Parts]
. . [Surfaces lenses BRL4V3NB]
. . . [Bloomed surfaces BRL4V3NC]

BRL 4V3 NE	. . Apertures
4V3 NH	. . Stops (optical equipment), diaphragms
4V3 NI	. . . Iris diaphragms
	. *Types of lenses*
4VC	. . Eyepieces, oculars
4VD	. . Thin lenses (general), astigmatic lenses
	. . . * Hypothetical, infinitely thin lens.
4VD L	. . . Cylindrical lenses
4VE	. . Thick lenses (general)
	. . . * Real lens, as distinct from infinitely thin lens.
4VF	. . Achromatic lenses
4VG	. . Apochromatic lenses
4VH	. . Contact lenses
4VI	. . Helical lenses
4VJ	. . Quadruple lenses
4VK	. . Spherical lenses
4VL	. . Aspherical lenses
4VM	. . Gas lenses
4VN	. . Zoom lenses
4VO	. . Refracting prisms
4VQ	. . Converging lenses
4VR	. . Diverging lenses
4VS	. . Complex lens systems
4VT	. . Geodesic lenses, integrated optics
4VU	. . Other forms of lens
	. . . * Arrange A/Z; eg zone plates BL5 VXZ.
4WB	Prisms
	. * See also Lenses & prisms (together) BRL 4UY
4WC	. Beam splitters
	. . * Arrangement of 3 prisms.
4WD	. Nicol prisms
	. . * See also Polarizers BRL 4WS
4WF	Reflectors
4WG	Gratings
4WM	Mirrors
4WN	. Schmidt corrector, Schmidt plates
4WP	. Plane mirrors
4WQ	. Spherical mirrors, curved mirrors
4WS	Polarizers
	. * See also Nicol prisms BRL 4WD
4WT	. Piles (optics)

[Optical properties of bulk matter BRL]
[Practical & experimental optics BRL36]
[Materials & equipment BRL3B]
. . [Optical instruments BRL4]
. . . [Optical components BRL45]
. . . . . [Lenses & prisms together BRL4UY]
. . . . . . [Polarizers BRL4WS]
. . . . [Piles optics BRL4WT]

BRL 4WU . . . . Kerr cells
4WY C . . . . . . Collimators, finders
4WY D . . . . Autocollimators
4WY F . . . . . . Focusing instruments
4WY H . . . . . . Optical scanners
4WY I . . . . Flying spot scanners
4WY K . . . . . . Isolators (light)
4X . . . *Components special to a type of optics*
. . . . * See particle optics imaging (BM7 I) for
. . . . example.
5 . . . *Types of optical instruments*
. . . . * For instruments serving a particular purpose,
. . . . see latter; eg Magnification - Microscopy
. . . . BL7 J.
62 Investigative techniques in optics
69 . Physical methods
. *By operations on the phenomena*
. . Control
72 . . . Optical control
. . . . * For control of specific properties, etc. see the
. . . . latter; eg Luminance - Control
. . . . BRL LR7 2..
73 . . Production
73K . . . Light sources
74G . . Detecting & indicating
74T . . Recording
74V . . . Scanning
76 . . Measurement
. . . * Measurement of particular phenomena goes
. . . with the phenomenon; eg Diffractometers
. . . BRL FQ7 64.
. . . * For Photometry, see BRL L76.
7A . . Testing & evaluation
7I . . Imaging
7IC N . . . Optical transfer function
7IC P . . . . Phase transfer function
7ID . . . Beam handling
7IE . . . . Focusing, beam focusing
7IE G . . . . . Beam trapping, optical self-focusing
. . . . . . * See also Thermal blooming BRL FCL R
7IE H . . . . . . Self-trapping
7IF . . . . Resolution
7IF H . . . . Modulation
7IF S . . . . Scanning
7IF T . . . . Projection

[Bulk matter physics BR]
[Optical properties of bulk matter BRL]
[Practical & experimental optics BRL36]
. [Investigative techniques in optics BRL62]
. . [By operations on the phenomena]
. . . [Imaging BRL7I]
. . . . [Beam handling BRL7ID]
. . . . . [Projection BRL7IFT]

BRL 7IF V . . . . . . Collimation
7IX . . . . Magnification
7J . . . . . Microscopy
7M . . . Spectroscopy
8M . Quantum optics
. . * For Non-linear optics, see BRL 9M; for
. . Coherent optics see BRL PF; for Lasers &
. . Masers, see B6K QM.
Non-linear systems
9M . Non-linear optics (general)
. . * Studies electro-optical effects of very intense
. . light beams and how the light interacts with
. . and is propagated through matter.
. . * See also specific properties, processes, etc.
. . treated as non-linear systems; e.g. Optical
. . phase conjugation BRL FEY; Stimulated
. . scattering BRL FTM.
. . * See also Quantum optics BRL 8M
9NP . Adaptive optics
AF Energy interactions & forms
BB . Energy
BB9 2D . . Distribution of energy
BB9 2DM . . . Spectral distribution of energy
. . . (optics)
BB9 2DN . . . . Equal energy spectrum
BB9 2DP . . . . . Relative spectral distribution,
. . . . . spectral density
DA . Kinematics
. . Velocity
EXJ . . Transients
EYB BF . . Radiant flux
. . . * For luminous flux, see Luminosity
. . . BRL LBB F.
. . Radiant intensity
EYB BI . . . Radiance
. . Wave motion

Physical optics (bulk matter)

[Energy interactions & forms BRAF]
[Electrical & magnetic properties BRGY]
[E-m waves by frequency & wavelength]
[Optical properties of bulk matter BRL]
[Energy interactions & forms BRLAF]
[Kinematics BRLDA]
[Wave motion]

BRL F	Physical optics (bulk matter), electromagnetic optics (bulk matter)
	. * Light as electromagnetic waves; for geometrical optics, see Rays BRL GC.
FC	Propagation, transmission
	. * For optical fibres, see Optical techniques BRL 3TF.
	. Coefficients
FC2 QK	. . Transmittance optics, transmission coefficient
	. . . * See also Opacity BRL FIL
FC4	. Instruments
FC4 UF	. . Filters
FCD C	. Velocity of propagations
FCI	. Transmissivity
FCL	. Amplification
	. . * For laser techniques, see B6K S
FCL R	. . Thermal blooming (nonlinear optics)
FCL S	. . Parametric oscillation amplification
FCM	Attenuation
FD	Frequency
FD7 T	. Conversion (optical frequency)
FD7 U	. . Optical harmonic generation
FDM	. Mixing frequency (non-linear optics)
FE	Spectra
FE7 3	. Production
	. . * For diffraction gratings, see BRL FQ4 PW.
FEX	Phase
FEY	. Optical phase conjugation
FF	Coherence
	. * For coherent light, see BRL PF.
	. Transients
FFC F	. . Optical coherent transients
FG	Emission
FGK	. Exitance, emittance
FGO	. . Luminous exitance
FGP	. . Radiant excitance, radiant emittance
FGQ	. Incandescence
FGQ S	. . Candoluminescence

[Optical properties of bulk matter BRL]
[Energy interactions & forms BRLAF]
[Kinematics BRLDA]
[Wave motion]
[Emission BRLFG]
. [Incandescence BRLFGQ]
. . [Candoluminescence BRLFGQS]

BRL FH	. Luminescence
	. . * Emission resulting from non-thermal processes.
	. . * For luminosity, see BRL L.
	. . Materials
FH3 TP	. . . Phosphors
	. . *By duration of luminescence*
FHL	. . . Fluorescence
	. . . . * Luminescence ceases on removal of energy source.
FHN	. . . Phosphorescence
	. . *By method of excitation*
FHP B	. . . Triboluminescence
	. . . . * By friction.
FHP H	. . . Sonoluminescence
FHQ	. . . Thermoluminescence
FHQ R	. . . . Radiothermoluminescence
FHR	. . . Electroluminescence
FHR T	. . . . Galvanoluminescence
FHR U	. . . . Cathodoluminescence
FHS	. . . Photoluminescence
FHS W	. . . . Radiophotoluminescence
FHT C	. . . Chemiluminescence
FHT E	. . . Bioluminescence
	. . . . * Optical aspects only.
FHV	Wavetrain
FI	Transparence
FIJ E	. Transparent bodies behaviour
FIJ G	. Self-induced transparence
FIJ H	. Non-transparence
	. . * For shadows, see BRL FTH.
FIK	Translucence
FIL	Opacity
	. * Reciprocal of transmittance BRL FC2 QK.
FIM	Visibility
FL	Absorption
	. Filters
FL4 UF	. . Absorption filters
FLQ	. Dichroism
	. . * Selective absorption by crystals in one plane and not another.
FLR	. . Pleochroism
	. . . * Colour effects of dichroism.
FLT	. Trichroism
FLX	Refraction & reflection (together)
FM	. Refraction
FM4	. . Instrumentation
FM4 UR	. . Rangefinders
FM9 BI	. . Refractive index
	. . . * See also Schlieren device B6L FQ

Optical properties of bulk matter

[Optical properties of bulk matter BRL]
[Energy interactions & forms BRLAF]
[Kinematics BRLDA]
[Wave motion]
[Refraction & reflection together BRLFLX]
. [Refraction BRLFM]
. . [Refractive index BRLFM9BI]

BRL FM9 BIP	. . . Becke line
FMJ	. . Double refraction, birefringence, photoelastic effect
	. . . * See also Double refraction techniques B6L KMJ
	. . . *By medium*
FMJ M	. . . . Circular birefringence
FMJ N	. . . . Elliptical birefringence
FMJ P	. . . . Stress birefringence, mechanical birefringence, photoelasticity
FML	. . . Extraordinary rays
FN	. Reflection
FN4	. . Instrumentation
FN4 UF	. . . Reflection filters
FNL	. . Total internal reflection
FNL 3R	. . . Materials
FNL 3TF	. . . . Optical fibres
FNM	. . Surface reflection
FNM R	. . . Fresnel reflection
FNQ	Images (optics)
	. * See also imaging as a technique, B7I
FNQ 37I	. Aberrations
FNQ 4	. Instrumentation
FNQ 4L	. . Converters (images)
FNQ 4SS	. . Detectors (images)
FNQ 4TU	. . Intensifiers (images)
FNQ 62	. Investigative techniques
FNQ 73P	. . Image processing & restoration
FNQ 73R	. . Reduction (images)
FNR	. Real images
FNS	. Virtual images
FNS T	. Stereoscopic images
	. . * For holography, see B7K.
FP	Polarization
FP7 3	. Production
FP7 3D	. . Rotation of plane of polarization
	. . . * For Kerr effect, see BRL JP.
FP7 3DR	. . . Rotatory power (polarization)
FP7 3DV	. . . Optical activity
	. . *By source*
FP7 3LE	. . . Irregular crystals
FP7 3LG	. . . Quartz
FP7 3LJ	. . . Other crystals
FP7 3LK	. . . Amorphous substances
FP7 3M	. . . Wave surfaces of refraction
FP7 3MN	. . . . Single-sheet surfaces
FP7 3MP	. . . . Double-sheet surface
	. . *Effects of particular media, structures*
FP7 3MR	. . . Ray axes, optic axes (polarization)
	. Measurement

[E-m waves by frequency & wavelength]
[Optical properties of bulk matter BRL]
[Energy interactions & forms BRLAF]
[Kinematics BRLDA]
[Wave motion]
[Polarization BRLFP]
. [Measurement]

BRL FP7 6	. . Polarimetry
FP7 64	. . . Polarimeters, polariscopes
	. . . . Instrumental outputs
FP7 64T	. . . . . Polaroids
	. *Properties*
FPQ B	. . Plane of polarity
FPQ D	. . Degree of polarization
FPQ E	. . . Depolarization
	. *Types of polarization*
	. . * For birefringence, see Double refraction BRL FMJ.
FPS	. . Plane polarization
FPT	. . Circular polarization
FPU	. . Elliptical polarization
FPV	. . Chromatic polarization
FQ	Diffraction
FQ4	. Instrumentation
FQ4 FWG	. . Diffraction gratings
FQ4 FWH	. . . Echelon gratings, echelles
FR	Interference
	. * For holography, see B7K.
	. Instrumentation
FR4	. . Interferometers
FR4 V	. . Interference filters
FRL	. Interference fringes
	. Interference patterns
FRN	. . Inclusion pattern interference
FRO	. . . Labradorescence
FT	Scattering
	. *Types of scattering by particle*
	. . * See note under Types of scattering at BFT.
	. . * Add to BRL FT letters J/M following BMF T if required.
FTH	. Shadows
FTM	. Stimulated scattering (general)
	. . * For laser techniques, see B6K S.
FTN	. Brillouin scattering
FTO	. Mie scattering
FTP	. Raman scattering
FTP W	. . Coherent antiStokes Raman scattering
FTQ	. Rayleigh scattering
FTR	. Tyndall effect
FTS	. *Types of scattering special to a particle*
	. . * For example, Coulomb scattering BNG FTS.
FUF	Diffusion
FUP	Dispersion

Optical properties of bulk matter

[Electrical & magnetic properties BRGY]
[E-m waves by frequency & wavelength]
[Optical properties of bulk matter BRL]
[Energy interactions & forms BRLAF]
[Kinematics BRLDA]
. [Wave motion]
. . [Dispersion BRLFUP]

. . . * For optical rotation, see BRL FP7 3D.
BRL FUP 4 . . . Instrumentation
FUP 4V . . . . Dispersion prisms
. . *Types of light motion*
FWR . . . Spherical waves
FWS . . . Cylindrical waves
FYN . . . Solitons
. . . Rays
GC . . . . Geometrical optics (bulk matter)
. . . . . * See also Optical instruments BRL 4
Interactions with other energy forms
GH . Acoustooptics, acoustic optics
. . * See also Brillouin scattering BRL FTN
GN . . Sonoluminescence
. . . * Produced by high-frequency sound waves.
GP . Thermooptics
H . Electrooptics
IU . . Stark effect
J . . Magneto-optical effects
. . . *Named effects*
JM . . . . Couton-Mouton effect (magneto-optics)
JN . . . . Faraday effect
JP . . . . Kerr effect, magneto-optical effect
JQ . . . . . Pockels effect
JS . . . . Voigt effect (magneto-optics)
JT . . . . Zeeman effect (magneto-optics)
M . *Relations with particles*
. . * Add to BRL M letters LX/Q following B if applicable.

[Energy interactions & forms BRAF]
[Electrical & magnetic properties BRGY]
[E-m waves by frequency & wavelength]
[Optical properties of bulk matter BRL]
[Energy interactions & forms BRLAF]
[Interactions with other energy forms]
. [Relations with particles BRLM]

Special optical properties & processes
BRL KY . Optical illusions
L . Luminosity
. . * Attribute of light sources which give visual sensation of brightness.
. . * See also Luminescence BRL FH
. . Measurement
L76 . . . Photometry (light)
L76 4 . . . . Photometers
L78 Q . . . . Physical photometry
. . . . . * Measurement by physical receptors, radiometers, etc.
L78 R . . . . Visual photometry, subjective photometry
. . . . . * Evaluation by visual effect in the eye of the observer.
. . Radiant flux
LBB F . . . Luminous flux, light flux
LBB G . . . . Illuminance, light flux density, illumination density
LBB I . . . . Luminous intensity, light intensity, candlepower
. . . . . Units
LBB I77 . . . . . . Candela
LFH R . . . Electroluminescence
LM . . Radiant quantities
LN . . Luminous quantities
LR . . Luminance, brightness
LT . . Shade
M . Colour
. . Measurement
M76 . . . Colorimetry
M78 Q . . . . Spectral methods (colour)
M78 R . . . . Comparison method (colour), empirical method (colour)
M8B . . Theory
M8V . . . *By specific theorists*
. . . . * A/Z by name; eg Goethe.
. . Frequency
MFD . . . Physical colour
. . . . *For physiological colour, see BRL MV.
MFN . . Reflection
MFN N . . . Surface colour
MH . . Electrochromism
. . Luminosity
ML . . . Brightness
. . *Special chromatic properties*

Special optical properties & processes

[Electrical & magnetic properties BRGY]
[E-m waves by frequency & wavelength]
[Optical properties of bulk matter BRL]
[Energy interactions & forms BRLAF]
[Special optical properties & processes]
[Colour BRLM]
[Special chromatic properties]

BRL MM . Chromaticity
. . * Objective description of the colour
. . quality of a visual stimulus
. . irrespective of its luminance.
. . Chromaticity & luminance together
. . completely specify a colour stimulus.
MM7 6 . . Measurement
MM7 6T . . Chromaticity coordinates
MM7 6V . . Tristimulus values
MN . Colour diagrams
MNP . . Two-dimensional colour
. . representation
MNR . . Three-dimensional colour
. . representation
MP . Chrominance
MQ . Saturation (colour), chroma
MR Composition of colours
. * For shade, see luminosity BRL LT.
Types of colours
MU . Colourlessness
MV . Physiological colour
. . * For physical colour, see
. . wavelength BRL MFD.
MW . . Hue
NB . Primary colours
NB2 M . . Colour equations
. . *Operations*
NC . . . Mixing colours (operation)
NE . . . . Additive mixing (colours)
. . . . . * Eg mixing coloured lights as
. . . . . in TV.
NF . . . . Subtractive mixing
. . . . (colours)
NFN . . . . . Colour filter transmission
NFP . . . . . Pigment mixing
NH . Secondary colours
NJ . Complementary colour,
. compensatory colour
. . * Giving white when mixed.
NK . Equivalent colours
. *By wavelength*
NMB . . Colour groups
. . . * For physical colour, see
. . . BRL MFD
NMD . . White
NME . . Grey
NMF . . Black
NMG . . Red group
NMH . . . Specific colours
. . . . * Eg carmine, pink.

[Optical properties of bulk matter BRL]
[Energy interactions & forms BRLAF]
[Special optical properties & processes]
. [Colour BRLM]
. . [Types of colours]
. . . [By wavelength]
. . . . [Red group BRLNMG]
. . . . . [Specific colours BRLNMH]

BRL NMJ . . . . Orange group
NMK . . . . . Specific colours
. . . . . . * Eg reddish orange, brown.
NML . . . . . . Yellow group
NMM . . . . . Specific colours
. . . . . . * Eg reddish yellow, cream.
NMN . . . . Green group
NMP . . . . . Specific colours
. . . . . . * Eg yellowish green, pure green.
NMQ . . . . Blue group
NMR . . . . . Specific colours
. . . . . . * Eg ice blue, turquoise.
NMS . . . . Indigo
NMT . . . . Violet group
NMV . . . . . Specific colours
. . . . . . * Eg bluish violet, purple.
NP . . Colour systems, chromatic systems
. . . * Representation of colours in terms of a
. . . specific set of coordinates (eg wavelength,
. . . luminosity).
NPQ . . . Objective colour systems
NPS . . . Subjective colour systems
NQ . . . Monochromatic systems
NR . . . Dichromatic systems
NS . . . Trichromatic systems
NT . . . . Physiological chromatic systems
. . . . . * See also physiological colour BRL MV;
. . . . . equivalent colours BRL NK
NU . . . . . Named systems
. . . . . . * Arrange A/Z; eg Abney, OSA.
NV . . . Tetrachromatic systems
NW . . . Thermochromatic systems
NX . . . Photochromatic systems
Other special optical properties
OB . Optical bistability
OC . Optical multistability
OD . Optical switching (non-linear optics)
OE . Optical properties of substances (general)
Types of light by property
. * Add to BRL P letters C/Y following BF;
. * Add to BRL PQ letters A/D following BG; eg
PF . Coherent light
. . * For lasers, see B6K KS.
PFF CK . . Light modulation
PM . Refracted light
PN . Reflected light
PP . Polarized light

[Bulk matter physics BR]
[Energy interactions & forms BRAF]
[Electrical & magnetic properties BRGY]
. [E-m waves by frequency & wavelength]
. . [Optical properties of bulk matter BRL]
. . . [Energy interactions & forms BRLAF]
. . . . [Types of light by property]
. . . . . [Polarized light BRLPP]

BRL PWC . . . . . Isotropic light
PWD . . . . . Anisotropic light
PX . . . . . Light pulses
PYH . . . . . Continuous light waves
QB . . . . . Beams light
. . . . *By source*
QN . . . . . Natural light
QP . . . . . Artificial light
. . . . *By frequency & wavelength*
U . . . . . Infrared radiation
UFE . . . . . . Spectra
UN . . . . . . Near infrared radiation
UR . . . . . . Far infrared radiation
V . . . . . Light, visible light
. . . . . . * See notes at BRL.
W . . . Ultraviolet radiation
WN . . . . Long wave ultraviolet radiation
WV . . . . Vacuum ultraviolet radiation, far ultraviolet radiation
X . . . X-rays, Roentgen rays
. . . . * Penetrating electro-magnetic radiation, usually generated by accelerating electrons to bombard a solid body, or by inner shell transition of atoms.
XN . . . . Grenz rays, Infra-Roentgen rays
XP . . . . Continuous X-rays, Bremsstrahlung
XQ . . . . . Cosmic ray X-rays
XT . . Gamma radiation
. . . * High energy photons, especially as emitted by a nucleus in a transition between two energy levels.
YU . . . Non-solar gamma radiation
YV . . H-alpha radiation, alpha radiation
. . . * See Nuclear reactions BOF KR

BRM Particle physics of bulk matter, high energy physics of bulk matter
. * Add to BRM letters M/Q following B; eg
BRM O . Nuclear physics of bulk matter

[Physics B]
[Bulk matter physics BR]
[Energy interactions & forms BRAF]
. [Particle physics of bulk matter BRM]
. . [Nuclear physics of bulk matter BRMO]

BRN States of matter, physical phases (states of matter), systems of bulk matter, forms of matter, phases
. * new file bu2.sch 27.6.97; 3.7.97; 22.7.97; 14.8.97 20.8.97; 12.11.97; 29.6.98; 31.7.98; 12.8.98; 27.8.98
. * For the effects of these on chemical behaviour, see Chemistry C.
. *Processes*
BRN 9HH . . Change of state of system
. . . * Use BRN P.
P . . Change of state, thermodynamic changes of state, phase changes
. . . * Process by which a substance changes from one state to another without a change in temperature; accompanied by change in volume and in degree of randomness in the internal structure.
. . . * Note that 'phase' is also used to represent a particular chemical substance in a mixture, or a particular crystal structure in a mixture of such structures. If these meanings arise in this class, the array (Systems by number of components) is provided, at BRR N. Here, 'phase' means a state of matter.
. . . Mathematics
P2M 9L . . . . Equations
. . . . . Named equations
P2M 9NB . . . . . Boyle-GayLussac equation
P2M 9NC . . . . . Clapeyron equation
P2M 9NV . . . . . Van der Waals equation
. . . *Conditions, parameters, influences*
P94 BC . . . . Degrees of freedom
P94 C . . . . Critical point
P94 E . . . . Volume
P94 F . . . . . Constant volume
P94 G . . . . . Decreasing volume
P94 H . . . . . Increasing volume
P94 J . . . . Pressure
P94 JC . . . . . Critical pressure
P94 K . . . . . Constant pressure
P94 L . . . . . Decreasing pressure
P94 M . . . . . Increasing pressure
P94 N . . . . . Other pressure conditions
. . . . . . * Add to BRN P94 N letters P/S following BSB J if applicable.
P94 O . . . . Velocity
P94 P . . . . Thermal conditions
P94 PC . . . . . Critical temperature
P94 Q . . . . . Constant temperature
P94 R . . . . . Adiabatic change of state
. . . . . Decreasing temperature
P94 S . . . . . . Cooling (change of state)

[States of matter BRN]
[Processes]
[Change of state BRNP]
[Conditions]
. [Thermal conditions BRNP94P]
. . [Decreasing temperature]
. . . [Cooling change of state BRNP94S]

. . . . * For heat loss in general, see BRG T.
BRN P94 T . . . . Supercooling
. . Increasing temperature
P94 V . . . Heating (change of state)
. . . . * For heat gain in general, see BRG U.
P94 W . . . . Superheating
P94 Y . Electro-magnetic conditions
P95 . . Electrical conditions
P96 . . Magnetic conditions
P97 . . Radiation conditions
R Phase transformations, phase transitions
. * For transitions in a particular state of matter, see the latter.
. * For order-disorder transformations, see solid solutions BVU SNR.
. Critical points
R94 E . . Critical state, transition points
S . Latent heat, specific latent heat
. . * Quantity of heat absorbed or released in thermal phase transformation. Change of internal energy of a physical system without change of temperature.
. . * See also Thermochemistry C
T . Phase equilibrium
TP . . Phase diagrams, constitution diagrams
TR . . Phase rule
TT . . Triple point, three-phase equilibrium
TY . Commensurate-incommensurate transformation
U . Allotropic transformation
. . * For allotropes, see Chemistry C
V . Isothermal transformations
BRO *Types of change by states involved*
. * This class is used only when qualifying a particular state of matter.
. * See the states of matter involved. In the case of two different states, the denser one is cited first; eg both liquid into solid and solid into liquid go under solids (see BVO L). Note that the same citation order applies to mixed states of matter (dispersions); see note at BRT.

[Physics B]
[Bulk matter physics BR]
[States of matter BRN]
[Processes]
. [Change of state BRNP]
. . [Types of change by states involved BRO]

Subsystems, parts
. * Each subsystem may be divided in the same way as a system. See Add instructions at BRR.
BRP W . Impurities (states of matter)
BRQ . Surfaces
BRQ 36 . . Practical & experimental
6LF Q . . . Diffraction techniques
6LF QR . . . . Low energy electron diffraction
. . . Interferometry
6LF R . . . . Surface interferometry
. . *Properties & processes*
. . . * For surface chemistry, see Physical chemistry C
9J . . . Structure
. . . . Profile, contours
9JV . . . . . Microtopography of surfaces
. . . Energy
BB . . . . Surface activity, surface energy
. . . . . * For surface tension, see BRQ BLX.
BBP . . . . . Potential energy
BBQ . . . . . . Potential barrier, potential hill
. . . . Quantum effects
BBQ 8O . . . . Tunnelling
BBQ T . . . . . Adhesion
BBQ X . . . . . Sorption physics
BBR . . . . . . Absorption
BBS . . . . . . Adsorption
. . . Tension
BLX . . . . Surface tension
BLX M . . . . . Wetting
BLX P . . . . . Angle of contact
CA . . . Friction
. . . . * See Solids - Surfaces BVQ CA
EY . . . Radiation properties
FGK . . . . Exitance
FNH . . . . Reflectance
FNI . . . . . Specular reflectance
FNJ . . . . . Diffuse reflectivity
FNK . . . . . Total relectivity
H . . . Electromagnetic properties
HWB . . . . Resistance
HWJ . . . . Resistivity
L . . . Optical properties
LFG O . . . . Luminous exitance
LFG P . . . . Radiant exitance, radiant emittance
M . . . Relations to particle physics
. . . . * Add to BRQ M letters M/Q following B.
O . . Interfaces
. . . * Surface separating two physical phases.
. . . Energy
OBB . . . . Interfacial energy
. . . . Surface tension

[Bulk matter physics BR]
[States of matter BRN]
[Subsystems]
. [Surfaces BRQ]
. . [Interfaces BRQO]
. . . [Energy]
. . . . [Surface tension]

BRQ OBL X . . . . . Interfacial surface tension
. . . . . . * For skin friction, see aerodynamic drag
. . . . . . BTB TKP.
OT . . . Contact surfaces
OU . . . Free surfaces
P . . . *Types by materials interfacing*
Q . Films (states of matter), surface films
. . * General studies only, covering solid & liquids
. . films & fluid molecular layers.
R . . Thin films
. . . Formation
RR . . . . Unctuosity, oiliness
RT . . . Molecular films
RU . . . . Monomolecular layers (films)
RV . . . Plateau figures
RW . . . Minimum surfaces, soap bubbles
SC . . Thick films
SG . . Coatings
T . . Membranes, porous media
TBJ . . . Pressure
. . . . * For osmotic pressure, see Liquids BUB JSO
TO . . . . Electrosmosis, electroendosmosis
BRR *Systems by state of matter*
. * Systems defined by particular physical properties
. (eg isotropic, low pressure, electrically conducting)
. are treated as indistinguishable from the state of
. matter qualified by the property; so dense liquids
. (say) are classified as Liquids - Density BUC L.
. * Each system may be divided as follows (where
. hyphen represents the clasmark of the system):
. * Add to - letters A/L following B;
. * Add to -M letters M/Q following B;
. * Add to - letters N/T following BR; eg
. Homogeneous systems - Mixtures BRS NRO.
. * Add to -U letters RU/V following BR for states
. dispersed in the medium (see note and example at
. BRT).
Systems by simplifying assumptions
. * Use only when the system is distinguishable from
. the properties qualifying the state of matter.
. * Add to BRR letters D/F following AYK; eg perfect
. gases BTR D.
BRR D . Ideal medium, perfect medium
Systems by number of components
. * Phases other than those represented by states of
. matter. See note at BRN P.
N . Single component systems
O . Mixtures (general)
. . * See note at BRT for citation order between the
. . states involved in a mixture, etc.

[Physics B]
[Bulk matter physics BR]
[States of matter BRN]
[Systems by number of components]
. [Mixtures general BRRO]

BRR S . . Solutions
. . . * See homogeneous systems BRS NRS
U . . Two component systems
V . . Three component systems
W . . Four component systems
X . . Five or more component systems
Systems by number of phases
. * Phase here means a state of matter, not a chemical
. substance or crystal structure.
. * For Allotropic systems, see Class C Chemistry.
BRS N . Homogeneous systems, single phase systems
NRO . . Mixtures
NRS . . . Solutions (general)
. . . . * Physical studies only; for solutions (critical
. . . . mixtures) see physical chemistry, C
T . Heterogeneous systems, inhomogeneous
. systems, multiple phase systems
TRO . . Mixtures
TU . . Anomalous systems
TV . . Binary systems (bulk matters)
TW . . Ternary systems (bulk matter)
TX . . Quaternary systems (bulk matter)
TY . . Five phases or more [
BRT . . Disperse systems, mixed states of matter,
. . dispersions
. . . * A system of particles dispersed and suspended
. . . in a solid, liquid or gas.
. . . * Physical considerations only; most of the
. . . literature goes in physical chemistry. If in doubt,
. . . prefer physical chemistry.
. . . * Mixtures, solutions, colloid dispersion, etc. of
. . . one state in another state go under the denser
. . . state; eg gases in liquids go under liquids.
. . . * Use -U to introduce states dispersed in a
. . . system; eg BVU G Solids - Dispersions in -
. . . Gas in solid.
BRT HBH . . . Electro-magnetic field
HDE . . . . Electrokinetic effect
HDF . . . . . Streaming potential
HDG . . . . . Sedimentation potential
HDH . . . . . Electrophoresis
. . . . . . * For electrosmosis, see Membranes
. . . . . . BRQ TO.
U . . . Suspensions
. . . . * See fluids BST U
V . . . Colloids
. . . . * Consist of ultramicroscopic particles,
. . . . intermediate between those of a true solute
. . . . and those of a suspension. Most substances
. . . . can be brought to the colloidal state by
. . . . suitable technique.
Y . . . Composite materials

[Physics B]
[Bulk matter physics BR]
[States of matter BRN]
[Systems by number of phases]
. [Heterogeneous systems BRST]
. . [Disperse systems BRT]
. . . [Composite materials BRTY]

. . . . * Displaying the distribution throughout one
. . . . material of another material in fine particle
. . . . form.

Systems by particular state of matter

BRU . Plasmas & fluids
. . * Add to BRU letters A/X following BS Fluids so
. . far as applicable.
. . Mechanics
BRU B . . . Flow of plasmas & fluids
B9B . . . . Dimensions of a physical quantity
. . . . . * Expressing the quantity in terms of powers
. . . . . of fundamental quantities (e.g. mass, length,
. . . . . time).
B9B K . . . . . Similarity parameters, dimensionless
. . . . . numbers
. . . . . . * Usually considered in relation to
. . . . . . aerodynamic and fluid flow. See BSB 9BL
. . . . . . for enumeration of particular numbers
. . . . . . (which may be added here as necessary).
GP . . Thermal properties
GW . . . Low temperature
. . Magnetism
J . . . Magnetohydrodynamics, MHD
. . . . * Behaviour of a conducting plasma or fluid in
. . . . a magnetic flux. The motion of the plasma
. . . . induces an electrical field which interacts with
. . . . the applied magnetic field to cause a change in
. . . . motion itself.

[Physics B]
[Bulk matter physics BR]
[States of matter BRN]
[Systems by particular state of matter]
[Plasmas & fluids BRU]
[Magnetism]
. [Magnetohydrodynamics BRUJ]

BRV Plasmas
. * A gas at a temperature high enough for
. most of its atoms to be ionized.
. * See also nuclear fusion BOX B
. * Add to BRV letters A/X following BS
. Fluids so far as applicable.
BRV 36 . *Practical & experimental*
69 . . Physical methods
6H . . . Electromagnetic
6J . . . . Toroidal systems (plasmas)
. . . . . Equipment
6J3 U . . . . . . Stellarators (plasmas)
. . Production
73 . . . Plasma generation, plasma
. . . sources
73K . . . . Formation of plasmas
73L . . . . Laser beam production of
. . . . plasmas
73L S . . . . Shock wave production of
. . . . plasmas
73M . . . Excitation (plasmas), ionization
. . . (plasmas), breakdown
. . . (plasmas)
73N . . . . Heating (plasmas), microwave
. . . . resonance of plasmas
73N K . . . . . Plasma focus
73P . . . . Acceleration (plasmas)
73P 76 . . . . . Measurement
73P 764 . . . . . . Plasma probes
73P 764 V . . . . Rogowski coil probes
73P 764 W . . . . Langmuir probe
73P 7M . . . . . Spectroscopy
. . *Special techniques*
73R . . . Containment, confinement,
. . . control
. . . . * Add to BRV 73 letters R/W
. . . . following BOX B7; eg
73R X . . . . Electrostatic confinement
73S . . . . Magnetic lines
73T . . . . Magnetic wells
73U . . . . . Linear field confinement
73U 3YB . . . . . . Magnetic traps, magnetic
. . . . . . bottles
73U 3YD . . . . Magnetic mirrors
73V . . . . . Pinch
73V T . . . . . . Toroidal field confinement
73V V . . . . Theta effect
73V W . . . . Tokamak confinement
73W G . . . . . Radiofrequency confinement
. . . . . (plasmas)

[States of matter BRN]
[Plasmas BRV]
[Practical & experimental BRV36]
. [Special techniques]
. . [Containment BRV73R]
. . . [Magnetic wells BRV73T]
. . . . [Radiofrequency confinement plasmas BRV73WG]

BRV 73W K . . . . Non-magnetic confinement
73W L . . . . Inertial confinement of plasmas
7F . Modelling
7FY T . . Heat-trap model

AF Plasma reactions
AF9 2J . Reaction rate
AG . Thermodynamics
AT . . Transport properties
AV . . . Diffusion in plasmas
AW . . . Viscosity
. Mechanics
B . . Flow of plasmas (general)
B8M . . . Quantum theory
BG . . . Forces
BJ . . . . Pressure
BJQ . . . . . High pressure plasmas
CB . . . . Elasticity
. . . . . * See also Vibration BRV E
CH . . . Statics
CL . . . . Density
CLP . . . . . Low density plasmas
. . . . . . * See also Collisionless plasmas BRV VL
CLT . . . . . High density plasmas
. . . . . . * See also Collisional plasmas BRV VR
CN . . . . Equilibrium
CR . . . . Instability
CRR B . . . . . Velocity-space instabilities
DA . . . Kinematics
DC . . . . Velocity
DC9 2D . . . . . Distribution
DE . . . Kinetics
. . . *Forms of motion*
. . . . Rotation
DN . . . . . Rotating plasmas
DS . . . . Periodic motion
E . . . . . Oscillation & vibration
F . . . . . Waves
FC . . . . . . Propagation
FPY . . . . . . Interactions
FPY W . . . . Plasma-wall interaction
FS . . . . . . Collisions
FS9 DR . . . . Atomic cross section
FYJ . . . . . . Linear waves
FYL . . . . . . Non-linear waves
FYN . . . . Solitons
GA . . . . . . Shock waves

[States of matter BRN]
[Plasmas BRV]
[Plasma reactions BRVAF]
. . . [Forms of motion]
. . . . [Periodic motion BRVDS]
. . . . . [Waves BRVF]
. . . . . . [Shock waves BRVGA]

. . . . *Special attributes & elements of flow*
. . . . . * Normal retroactive synthesis from BRG F/BRL is interrupted here in order to accommodate the large vocabulary of fluid flow. It is resumed at BRV M.
BRV GL . . . . . Jets
GLV . . . . . Arcs
GM . . . . . Drift waves, drift current
M *Other energy interactions & forms*
. * Normal retroactive synthesis from BR is resumed here after its interruption beginning at BRV F.
. * Add to BRV M letters GF/L following BR; eg
. Acoustic properties
MGH . . Plasma sound waves, magnetosonic waves
MGP . Thermal properties
MGP Q . . Heat
MGQ . . Heat transfer
MGV . . Temperature
. . Thermal regimes
MGW . . . Low temperature plasmas
. . . . * 10,000 K - 100,000 K
. . . . * See also Collisional plasmas BRV VR
MGW W . . . . Thermal plasmas
MGW X . . . . Cold plasmas
MGX . . . High temperature plasmas
. . . . * Over 100,000 K.
. . . . * See also Collisionless plasmas BRV VL
MGY . Electrical & magnetic properties
MHL . . Electric potential
MHN . . Electrostatic waves
MHR . . Discharges
MHU . . Conductivity
MHW B . . Resistance
MIS . . Dielectric properties
. . Magnetic properties
. . . Magnetohydrodynamics
MJ . . . . Magnetoplasmadynamics
. . . . . * Behaviour of a conducting fluid (ionized gas, plasma, etc.) under the influence of a magnetic flux.
. . . . . Waves
MJF . . . . . . Alfven waves, hydromagnetic waves, plasma magnetohydrodynamic waves

[States of matter BRN]
[Plasmas BRV]
[Other energy interactions & forms BRVM]
. [Electrical & magnetic properties BRVMGY]
. . [Magnetic properties]
. . . . . [Waves]
. . . . . . [Alfven waves BRVMJF]

BRV MK	. . Radiation
MKF H	. . . Emission in plasmas
MKF L	. . . Absorption in plasmas
MKF T	. . . Scattering in plasmas
ML	. . . Optical properties
MM	. Particle physics of plasmas
MMD MW	. . Particle orbits in plasmas
MMG B	. . Particle beam interaction in plasmas
MNH	. . Charged particles (interaction with plasmas)
MNP	. . Electron physics of plasmas
MNP BD	. . . Electron state (plasmas)
MQU	. . Ion physics of plasmas
	. . . Dissociation
	. . . . Decay
MQU FJ	. . . . . Plasma loss
MQU FJ9 CI	. . . . . Plasma lifetime
	. Change of state
NP	. . Natural formation of plasmas
	. . . * See also plasma production techniques BRV 73
	Subsystems, parts of plasmas
PW	. Impurities in plasmas
	Systems by number of components
RN	. Single component systems
	Systems by number of phases
SN	. Homogeneous plasmas
ST	. Inhomogeneous plasmas
U	. *Systems by phase media*
	. . * For plasmas in other states (gases, liquids, solids), see the other state; eg gaseous plasmas BTU B.
	Special types of plasmas
	. *By form, shape*
VB	. . Plasmoids
VC	. . . Clouds (plasmas)
VD	. . . Rings (plasmas)
VE	. . Columns (plasmas)
VF	. . Layers (plasmas)
VG	. . . Sheaths (plasmas), plasma double layers
VH	. . Slab plasmas
VJ	. . Cylinders (plasmas)
	. *By activity*
VL	. . Collisionless plasmas
	. . . * See also Low density plasma BRV CLP
VM	. . . Non-equilibrium plasmas
VN	. . . Non-uniform plasmas
VP	. . . Relativistic plasmas
VR	. . Collisional plasmas

[States of matter BRN]
[Systems by particular state of matter]
[Plasmas & fluids BRU]
[Plasmas BRV]
. [Special types of plasmas]
. . [By activity]
. . . [Collisional plasmas BRVVR]

	. . . . * See also High density plasma BRV CLT
	. . *By composition*
	. . . * For impure plasmas, see BRV PW
	. . . *By particle*
	. . . . * See Particle physics of plasmas BRV MM; eg Electron plasmas BRV MNP.
BRX	. . . *By chemical characteristics*
	. . . . * Alternative (not recommended) to locating in Class C Chemuistry; if this option is taken:
	. . . . * Add to BRX letters G/R following C; eg (notation is provisional)
BRX IC	. . . . Metallic plasmas
ICP	. . . . . Alkali metal plasmas
V	. . *By location*
	. . . * Alternative (not recommended) to locating these in their context. If this option is taken, proceed proceed as follows:
	. . . * Add to BRX V letters A/G following D; eg Cosmic plasmas, Geophysical plasmas.
BS	Fluids
	. * Many concepts included in this class are applicable mainly (and sometimes perhaps exclusively) to a particular type of fluid or plasma. For economy in scheduling and ease of reference and assignment of notation, all generally applicable concepts are given here, under fluids in general. Concepts which are largely or exclusively applicable only to a particular form of fluid are given under that form; eg lift in aerodynamics (BTB SB). But any concept appearing in any part of this general fluids class may be used in any type of fluid or plasma, and vice versa, should it prove to be applicable.
	. * Normal retroactive notation is modified for fluids in order to insert a large expansion under flow (beginning at BSG E). The add instructions are summarized here for convenience although they are reinforced at necessary points of the schedule.
	. * Add to BS letters A/GD following B;
	. * Add to BSM letters GF/L following BR;
	. * Add to BSM letters M/Q following B;
	. * Add to BS letters N/T following BR.
BS8 EK	. Kinetic theory
	. Quantum theory
M	. . Quantum fluids
BSA G	. Thermodynamics
	. . * For thermal properties narrowly, see BSM GP.
T	. . Transport properties
V	. . . Diffusion
W	. . . Viscosity, fluidity
WC	. . . . Dynamic viscosity

Fluids

[States of matter BRN]
[Plasmas & fluids BRU]
[Fluids BS]
[Thermodynamics BSAG]
. [Transport properties BSAT]
. . [Viscosity BSAW]
. . . [Dynamic viscosity BSAWC]

BSA WD . . . Kinematic viscosity
WI . . . Intrinsic viscosity
Mechanics of fluids
BSB . Flow of fluids (general)
BSB 36 . . Practical & experimental study
3U . . . Equipment
3YE . . . . Tunnels & tubes, rigs
. . . . . * Add to BSB 3Y letters E/U following BTB 3Y so far as applicable; eg BSB 3YS Expansion tubes.
76 . . . Measurement
764 . . . . Instruments
764 V . . . . . Venturi meters
764 W . . . . . Pitot tubes
8B . . Theory
8ES . . . Molecular theory of flow
9B . . Dimensions of a physical quantity
9BK . . . Similarity parameters, dimensionless numbers
9BL E . . . . Euler number
9BL F . . . . Froude number
9BL H . . . . Hartmann number
9BL K . . . . Knudsen number
9BM . . . . Mach number
9BM P . . . . . Critical Mach number
9BM V . . . . . Mach number of divergence
9BN . . . . Nusselt number
9BN P . . . . Prandtl number
9BO . . . . Reynolds number
9BP . . . . Others
. . . . . * Arrange A/Z.
G . . Forces
. . . * Many of the concepts in this class are more usually found under gases; see BTB G/BTC D for more details (which may be used here if required).
IM . . . Moments
IQ . . . . Pitching moments
IR . . . . Rolling moments
IS . . . . Yawing moments
IT . . . . Hinge moments
J . . . Pressure
. . . . *Properties & processes*
J92 D . . . . . Distribution of pressure
J94 . . . . . Pressure conditions
J94 E . . . . . . Critical pressure
J94 K . . . . . . Constant pressure
JBH . . . . . Pressure fields
JMP . . . . . Pressure effects
. . . . . . * See also Cavitation BSG NV

[Fluids BS]
[Mechanics of fluids]
[Flow of fluids general BSB]
[Forces BSBG]
. [Pressure BSBJ]
. . [Properties & processes]
. . . [Pressure effects BSBJMP]

BSB JN . . . . Centre of pressure
JO . . . . Head of pressure
JO9 2J . . . . . Gradients
JOS . . . . . Pressure drop, pressure loss
JOT . . . . . Pressure recovery
JOU . . . . . Pressure rise, pressure gain
. . *Types of pressure*
. . . * See also Osmotic pressure BUB JSO
JP . . . Low pressure
JQ . . . High pressure
JSB . . . Partial pressure
JSC . . . Static pressure
JSD . . . Dynamic pressure
JSF . . . Induced pressure
JSI . . . Impact pressure
JSJ . . . Base pressure
JSK . . . Kink pressure
JSN . . . Stagnation pressure
JW . External forces
K . . Deformation
L . . . Stress
Q . . Loading, loads
. . . * See also Dynamic stability BSC Q
. . . * Add to BSB letters Q/T following BTB so far as applicable. A selection of examples is given to show their scope.
QS . . . Compressibility
QU . . . Buffeting loads
QV . . . Gust loads
QW . . . Impact loads hydrodynamics
SB . . . Lift, normal force lift
TB . . . Drag
TS . . . Lateral force
TT . . . Drift
BSC B . Elasticity
. . * See also Vibration BSD V
D . Internal forces
H Statics
L . Density
N . Equilibrium
P . Stability
PS . . Static stability
Q . . Dynamic stability
. . . * Add to BSC Q letters T/V following BTC Q; eg longitudinal stability BSC QU.
R . Instability
X Dynamics
BSD A Kinematics
B . Displacement
C . Velocity

[Systems by particular state of matter]
[Plasmas & fluids BRU]
[Fluids BS]
[Mechanics of fluids]
[Flow of fluids general BSB]
[Kinematics BSDA]
. [Velocity BSDC]

. . * For types of flow defined by speed, see
. . BSJ R.
BSD C76 . . Measurement
C76 4 . . . Rate flowmeters
C92 J . . Gradient
CQ . . Mass flowrate
CR . . Internal velocity
CS . . Surface velocity
D . Acceleration
E Kinetics
. * For kinetic theory of fluids, see BS8 EK.
K Forms of motion
N . Rotational flow
P . . Vortices, vortex motion, vortex flow
. . . * See also cavitation BSG NV
PR . . . Breakdown of vortices
. . . *Properties*
PT . . . . Vorticity
. . . *Elements*
QF . . . . Vortex filaments
QH . . . . Vortex streets
QJ . . . . Vortex sheets
. . . *Types*
QL . . . . Bound vortices
QN . . . . Trailing vortices
QP . . . . Free axis vortex motion
QR . . . . Three-dimensional vortex
. . . . motion

S . Periodic motion
. . * See also Elasticity BSB C
BSE . . Vibration & oscillation
. . . * See also special forms under
. . . aerodynamics BTE; eg Flutter BTE H.
BSF . . Waves
BSF YR . . . Transverse waves
YU . . . Longitudinal waves
BSG A . . . Shock waves, compression waves
. . . . * See also Supersonic flow (gases)
. . . . BTJ W
AF . . . . *Properties & processes*
AFC M . . . . . Attenuation
. . . . *Types*
AG . . . . . Mach cones
. . . . . . * See also Sonic boom BTM GHT
AH . . . . . Attached shock waves
AJ . . . . . Detached shock waves
AL . . . . . Normal shock waves
AN . . . . . Oblique shock waves
AP . . . . . Standoff waves

[Fluids BS]
[Flow of fluids general BSB]
[Forms of motion BSDK]
. . [Waves BSF]
. . . [Shock waves BSGA]
. . . . [Types]
. . . . . [Standoff waves BSGAP]

BSG AR . . . Blast waves
AS . . . . Explosion waves
AT . . . . Denotation waves
AX . . . Expansion waves
D . . . Other types of waves
. . . . * Arrange A/Z; eg flexural waves
Special attributes & elements of flow
. * See note at BT on the applicability of the concepts
. below to different types of fluids.
. * Normal retroactive synthesis from BRG H/BRL is
. interrupted here in order to accommodate the large
. vocabulary of fluid flow. It is resumed at BSM.
ER . Flow regime
. . * Complete flow field.
EV . . Homogeneity (flow)
FC . Suction
FE . Aeronautical factors
. . * See BTG FE and the note there.
FT . Interference (fluid dynamics), interaction
. (fluid dynamics)
FV . Stagnation
FX . Choking
G . Aerodynamic heating
GH . Downwash
GJ . Upwash
GL . Sidewash
GP . Wake
GQ . Slipstream
GT . Turbulence
GV . Surging
GX . Ground effect, ground resonance
JB . Hydrodynamic characteristics
JD . . Spray
JF . Concurrent streams
JH . Mainstream
JJ . Downstream
JM . Streamtubes
JP . Streamlines
JQ . Sources
JS . Sinks
JT . Doublets
L . Jets
LM . . Jet streams
LN . . Mixing jets
LP . . Propulsive jets
LR . . Wall jets
LT . . Plumes
LU . . Couette flow
LV . Arcs (fluid flow)
NC . Leaks

Special attributes & elements of flow

[Systems by particular state of matter]
[Plasmas & fluids BRU]
[Fluids BS]
[Mechanics of fluids]
[Flow of fluids general BSB]
[Special attributes & elements of flow]
. [Leaks BSGNC]

BSG NG . Eddies
NJ . Surges
NL . Sloshing
NN . Circulation
NP . . Supercirculation
NR . Ripples
NT . Drops
NV . Cavities, cavitation
NX . . Bubbles
P . Boundary layer
. . *Processes*
PH . . . Transition (boundary layer)
PK . . . Separation boundary layer
PL . . . Reattachment
PM . . . Thickening boundary layer
PN . . . Growth boundary layer
PP . . . Interference boundary layer
. . *Types*
PS . . . Wall boundary layer
PT . . . Oscillating boundary layer
Q . . . Other
. . . . * Add to BSG Q letters DF/L
. . . . following BT; eg Laminar boundary
. . . . layer BSG QI.
Types of flow
. *By shape*
RC . . Conical flow
. *By compressibility*
RE . . Compressible flow
RG . . Incompressible flow
. *By viscosity*
. . * For free stream flow, see BSH W.
RJ . . Viscous flow
RJL . . . Slow viscous flow
RJP . . . Creeping viscous flow
RK . . Non-viscous flow, inviscid flow
RM . . Newtonian flow
RN . . Non-Newtonian flow
. . . * See also Multiphase flow BSG RP
RNL . . . Relativistic flow
RNV . . Forced flow
. *By phase*
RO . . Single phase flow
RP . . Multiphase flow
. . . * See also Non-Newtonian flow
. . . BSG RN
RQ . . . Two phase flow
RR . . . . Piston flow, plug flow, slug flow
RS . . . Three or more phases
. *By dimension*

[Systems by particular state of matter]
[Plasmas & fluids BRU]
[Fluids BS]
[Mechanics of fluids]
[Flow of fluids general BSB]
[Types of flow]
[By dimension]

BSG RT . One-dimensional flow
RU . Two-dimensional flow
RV . Three-dimensional flow
By symmetry
SE . Symmetrical flow
SF . Axisyymetrical flow
By equilibrium
SH . Equilibrium flow
SJ . Non-equilibrium flow, reactive
. flow, radiative flow
By degree of attachment
SM . Separated flow
SN . Unseparated flow
SR . Reattached flow
SS . Mixed flow
ST . . Re-energized flow
By stress
SV . Cavity flow
T . . Shear flow
By development
TQ . Secondary flow
TR . Fully developed flow
TT . Accelerated flow
TV . Decelerating flow
By pressure
. * For compressible & incompressible
. flow, see BSG RE; for molecular flow
. (Knudsen flow) see Gases BTG V.
U . Isobaric flow
UR . Rarefied flow
. . * Add to BSG letters UR/X
. . following BTG so far as applicable;
. . eg Molecular flow BSG V.
Y . Isothermal flow
By parameter changes
BSH B . Reversible flow
D . Irreversible flow
F . Adiabatic flow
H . Diabatic flow
I . Isentropic flow
J . Non-isentropic flow
By presence of velocity potential
L . Potential flow
P . Irrotational flow
. . * For rotational flow, see BSD N.
By continuity
S . Steady flow, continuous flow
U . Unsteady flow, transient flow
By degree of mixing
W . Free stream flow

Fluids

[Plasmas & fluids BRU]
[Fluids BS]
[Mechanics of fluids]
[Flow of fluids general BSB]
[Types of flow]
[By degree of mixing]
. [Free stream flow BSHW]

BSI	. Laminar flow, streamline flow, Poiseuille flow
BSI S	. Superlaminar flow
T	. Transitional flow
BSJ	. Turbulent flow
BSJ P	. Pulsating flow
	By speed
	. * Add to BSJ letters R/X following BTJ so far as applicable; eg
R	. Low speed flow
S	. Medium speed flow
T	. High speed flow
	Flow defined by bodies in fluid
BSK B	. Bodies in fluid flow
BBG	. . Control forces (fluid flow)
	. . . * On bodies in fluids.
BBG L	. . . Longitudinal controls
BBG N	. . . Lateral controls
BBG P	. . . Directional controls
	. . *Parts of bodies in flow*
D	. . . Surfaces
DL	. . . Corners, kinks
DN	. . . Bends
DP	. . . . Irregular bends
DR	. . . Holes
E	. . . . Orifices, apertures, openings
EL	. . . Constrictions
EN	. . . Widenings
EP	. . . Entries
EQ	. . . Exits
ER	. . . Forebodies
ET	. . . Centrebodies
EV	. . . Afterbodies
	. . *Types of bodies in flow*
F	. . . Streamlined bodies
G	. . . One-dimensional bodies (fluid flow)
H	. . . Two-dimensional bodies (fluid flow)
I	. . . . Porous media (bodies in flow), membrane flow
J	. . . . Manifolds
K	. . . . Nozzle
L	. . . . Conduits, ducts
LL	. . . . . Low gradient conduits
LP	. . . . . High gradient conduits
M	. . . . . Open conduits
N	. . . . . . Channels, passages, channel flow
NP	. . . . Semi-infinite channels

[States of matter BRN]
[Fluids BS]
[Mechanics of fluids]
. . . . . [Types of bodies in flow]
. . . . . . [Two-dimensional bodies fluid flow BSKH]
. . . . [Conduits BSKL]
. . . . [Open conduits BSKM]
. . . . [Channels BSKN]
. . . . [Semi-infinite channels BSKNP]

BSK O	. . . . Sluices
	. . . . * Channels allowing sudden rushes of fluid.
P	. . . . Closed conduits
PP	. . . . Pipes
PR	. . . . Networks
PX	. . . . Diffusers
Q	. . . . Three-dimensional bodies (fluid flow)
QL	. . . . Cubes
QP	. . . . Prisms
QR	. . . . Pyramids
R	. . . . Wedges
S	. . . . Bodies of revolution
T	. . . . Cylinders
V	. . . . Hollow cylinders, tubular cylinders
W	. . . . Discs
BSL C	. . . . Cones
D	. . . . Truncated cones
H	. . . . Spheres, balls
I	. . . . Cocentric spheres, concentric flow
	. . . . * See also Couette flow BSG LT
J	. . . . Hemispheres
L	. . . . Ellipsoids
N	. . . . Ogives
O	. . . . Toruses
P	. . . . Paraboloids
R	. . . . Rings
S	. . . . Boat-tail bodies
T	. . . . Blunt bodies
U	. . . . Bluff bodies
W	. . . . . . Other bodies in flow
BSM	*Other energy interactions & forms*
	. * Normal retroactive synthesis from BR is resumed here after its interruption beginning at BSG ER.
	. * Add to BSM letters GF/L following BR: eg
BSM GH	. Acoustic properties (fluids)
GP	. Thermal properties (fluids)
GQV	. . Convection
GQV S	. . . Natural convection, free convection
GQV T	. . . Forced convection
GWX	. . Zero temperature
	. . . * For superfluids, see BSM PBF.
H	. Electro-magnetic properties
	. . Conduction
HU	. . . Electrically conducting fluids

[States of matter BRN]
[Plasmas & fluids BRU]
[Fluids BS]
[Other energy interactions & forms BSM]
. [Electro-magnetic properties BSMH]
. . [Conduction]
. . . [Electrically conducting fluids BSMHU]

. . . . * For magnetohydrodynamics, see BRU J.
BSM M . *Relations with particle & atomic physics*
. . * Add to BSM letters M/Q following B; eg
P . . Atomic aspects
PBE G . . . Lowest energy state
. . . . * For superfluidity, see BSM PBF; for
. . . . absolute zero temperature, see BRG WX.
PBF . . . Superfluidity, superfluids
. . . . * State exhibiting apparently frictionless flow.
PBF B . . . . Superfluid flow
BSN States of matter
BSN P . Change of state
R . . Transitions
BSO . . *Types of changes of state*
. . . * See note at BRO.
BSP R . . . *Properties special to a type*
Subsystems, parts of fluids
W . Impurities
BSQ . Surfaces
BSQ GJH . . Mainstream
O . . Interfaces
Q . . Films
U . . Free surfaces
Types of fluids
. * See also properties defining quasi-types; eg
. Electrically conducting fluids BSM HU
. *By simplifying assumptions*
BSR D . . Perfect fluids, ideal fluids
F . . Real fluids
FR . . . Rarefied fluids
. *By number of components*
O . . Mixtures
. *By number of phases*
BSS N . . Homogeneous
T . . Heterogeneous systems
TV . . . Binary systems
. . . Anomalous fluids
U . . . . Non-Newtonian fluids, anomalous
. . . . viscosity
. . . . . * Fluids with two or more phases present at
. . . . . the same time.
. . . Mixed systems, dispersions
BST . . . . Fluid dispersions
. . . . . * For dispersions involving two states, cite
. . . . . first the denser state; eg fluids in solids go
. . . . . under solids.
BST U . . . . . Suspensions
. . . . . . * Systems in which the denser particles are
. . . . . . distributed through the less dense fluid.
BSU . . . . *By state dispersed*

[Plasmas & fluids BRU]
[Fluids BS]
[Types of fluids]
[By number of phases]
. [Heterogeneous systems BSST]
. . [Mixed systems]
. . . [By state dispersed BSU]

. . . . * See note at BRT re citation order.
BT Gases
. * Works relating to physics of air are treated as
. general works on gas physics unless their purpose
. is specifically to distinguish problems of air from
. those of other gases, in which case prefer BTV.
. * Many concepts associated primarily with gas
. physics (eg many types of flow) appear in the
. general class for fluids (BS). All concepts in fluid
. physics which are applicable to gases may be
. added to the latter if necessary (eg BTG FH
. Balancing (aerodynamics)). Similarly, if any
. concept enumerated under gases but not under
. fluids or liquids proves to be applicable to the
. latter, it may be added to them.
. * Add to BT letters A/U following BS; eg
BT8 EK . Kinetic theory
. Spatial properties
. . Anisotropy
BT9 DEL . . . Anisotropic gases
BTA G . Thermodynamics
T . . Transport properties
V . . . Diffusion
VT . . . . Penetration of gases, inflow (gases)
W . . . Viscosity
WS . . . . Internal friction (gases)
WT . . . . . Outflow
. Mechanics
BTB . . Pneumatics, flow of gases (general),
. . aerodynamics
BTB 36 . . . Practical & experimental study
3U . . . . Equipment
3XT . . . . . Physical models
3YB . . . . . . Simulators (gas flow)
3YE . . . . . Tunnels & tubes, rigs
. . . . . . *Parts*
3YF . . . . Special components
. . . . * Arrange A/Z; eg nozzles
. . . . BTB 3YF N.
3YG . . . . Walls
. . . . . . *Types*
3YJ . . . . Rotating equipment (gas
. . . . physics)
. . . . * For spinning towers, see
. . . . BTB 3YO V
3YL . . . . Wind tunnels
3YL 3BD . . . . Design & construction
. . . . *Types*

[Gases BT]
[Pneumatics BTB]
[Practical & experimental study BTB36]
. [Equipment BTB3U]
. . [Tunnels & tubes BTB3YE]
. . . [Types]
. . . . [Wind tunnels BTB3YL]
. . . . . [Types]

BTB 3YM . . . . . . *By flow properties*
. . . . * Add to BTB 3YM letters D/M following BT (except for JR/JX).
3YN . . . . . . *By speed*
. . . . * Add to BTB 3YN letters R/X following BTJ; eg transonic wind tunnels BTB 3YN V.
3YO V . . . . . . Vertical wind tunnels, spinning towers
3YP . . . . Shock tubes
3YQ . . . . Shock tunnels
3YS . . . . Expansion tubes
3YU . . *Special to a situation*
. . . * E.g. electrolytic tanks (potential flow) BTH L3Y U
76 . Measurement
9B Dimensions of a physical quantitity
9B2 X . Dimensional analysis
9BK . Similarity parameters
9BL K . . Knudsen number
9BM . . Mach number
9BO . . Reynolds number
G Forces
IM . Moments
IQ . . Pitching moments
IR . . Rolling moments
IS . . Yawing moments
IT . . Hinge moments
IX . Attitude
. . * Angle of aerofoil to flow level.
J . Pressure
. . *Properties & processes*
J94 . . . Pressure conditions
J94 E . . . . Critical pressure
J94 K . . . . Constant pressure, isobaric
JBH . . . Pressure fields
JMP . . . Pressure effects
JN . . . Centre of pressure
. . *Types of pressure*
JP . . . Low pressure
JQ . . . High pressure
JSB . . . Partial pressure
JSC . . . Static pressure
JSD . . . Dynamic pressure
JW . External forces
Q . . Loading, loads
QS . . . Compressibility
QU . . . Buffeting loads
QV . . . Gust loads

[Gases BT]
[Mechanics]
[Pneumatics BTB]
[Forces BTBG]
. [External forces BTBJW]
. . [Loading BTBQ]
. . . [Gust loads BTBQV]

BTB QW . . . Impact loads
SB . . . Lift, normal lift
SB2 QK . . . . Lift coefficient
SB9 2D . . . . Lift distribution
SCK . . . . Kussmer function
SF . . . . Lift vector
SH . . . . Lift slope
SJ . . . . Interference (lift)
SL . . . . Lift effects
SN . . . . Non-linear lift
SP . . . . Maximum lift
TB . . . Drag (aerodynamics), air resistance (aerodynamics)
TB2 QK . . . . Drag coefficient
TF . . . . Drag rise
TG . . . . Drag divergence
TH . . . . Foredrag
TJ . . . . Profile drag
TK . . . . . Form drag
TKP . . . . . . Surface friction drag, skin friction drag
TKR . . . . Wave drag
TL . . . . Induced drag, lift-induced drag
TN . . . . Parasitic drag, interference drag
TP . . . . Coulomb drag
TQ . . . . Zero lift drag
TS . . . Lateral force, side force
TT . . . Drift
TU . . . Wind loading
TV . . . . Gust loading
TW . . . . Thrust
TX . . . . . Vertical thrust
UN . . . Torsion
VE . . . Elastic deformation
BTC B . . Aeroelasticity (general)
. . . * Interaction of aerodynamic forces & the elastic reactions of bodies.
. . . * See also Vibration BTE
. . . Viscoelasticity
C . . . . Viscoelastic gases
D . Internal forces
H Statics
K . Weight
KN . . Buoyancy
. . . * See also vertical thrust BTB TX
L . Density
N . Equilibrium
P . Stability
PS . . Static stability
Q . . Dynamic stability

[Types of fluids]
[Gases BT]
[Mechanics]
[Pneumatics BTB]
[Statics BTCH]
. [Stability BTCP]
. . [Dynamic stability BTCQ]

	. . . * See also Oscillation BTE
BTC QTR	. . . Stability derivatives
QU	. . . Longitudinal stability
QUH	. . . . Short-period longitudinal stability
QUP	. . . . Phugoid stability
QV	. . . Lateral stability, directional stability
QVK	. . . . Snaking
	. . . . . * Uncontrolled oscillation in yaw.
QVR	. . . . Dutch rolling
R	. Instability
	Dynamics
	. * For aerodynamics, see gas mechanics in general (BTB).
XN	. Non-stationary gas dynamics
BTD A	. Kinematics
B	. Displacement
C	. . Velocity
	. . . * For types of flow defined by speed, see BTJ R.
	. . . Measurement
C76	. . . . Anemometry
D	. . Acceleration
E	. Kinetics
	. . * For kinetic theory of gases, see BT8 EK.
E8G	. . Brownian movement
EG	. . Mean energy of molecules, mean velocity of molecules
	. *Forms of motion*
N	. . Rotational flow
P	. . . Vortices, vortex mmotion, vortex flow
S	. . Periodic motion
V	. . . Vibration aeroelasticity
	. . . . * Use BTE.
BTE	. . . Vibration & oscillation
	. . . . * See also aerolasticity BTC B
	. . . . * Add to BTE letters A/E following BE.
	. . . . * Add to BTE F letters F/U following BE so far as applicable.
	. . . . *Special forms*
BTE H	. . . . . Flutter
	. . . . . . * Sustained oscillation of body.
J	. . . . . . Flexure-torsion flutter
L	. . . . . Buzz
	. . . . . . * Severe vibration, especially of control surfaces.
N	. . . . . Buffeting (oscillation)
	. . . . . . * Irregular oscillation of body.
BTF	. . . Waves
BTF YU	. . . . Longitudinal waves

[Gases BT]
[Pneumatics BTB]
[Dynamics]
. [Forms of motion]
. . [Periodic motion BTDS]
. . . [Waves BTF]
. . . . [Longitudinal waves BTFYU]

BTG A	. . . . Shock waves, compression waves
	. . . . . * See also Sonic boom BTM GHT; Supersonic flow BTJ W
	. . . . . *Properties & processes*
AFC M	. . . . . . Attenuation
	. . . . . *Types*
AG	. . . . . . Mach cones
AH	. . . . . . Attached shock waves
AP	. . . . . . Standoff waves
AR	. . . . Blast waves
AX	. . . . Expansion waves
D	. . . . Other types of waves
	Special attributes & elements of flow
	. * Normal retroactive synthesis from BRG H/BRL is interrupted here in order to accommodate the large vocabulary of fluid flow. It is resumed at BTM
	. * The main enumeration of the special properties and types of flow is given under fluids in general, at BSG. A number of the concepts prominent in gas dynamics are repeated below for convenience and to demonstrate the scope of the synthesis. But all the classes under fluids may be used under gases if required.
	. * Add to BT letters GE/L following BS; eg
ER	. Flow regime
FC	. Suction
FE	. Aeronautical factors (airflow)
	. . * Alternative (not recommended) to locating in Technology U.
FG	. . Blowing
FH	. . . Balancing (aerodynamics)
	. . . . * Minimizing control forces needed to balance air loads.
FK	. . Rotational field (aerodynamics), curl (aerodynamics)
FL	. . Autorotation, windmilling
FM	. . . Spin (aerodynamics)
FN	. . Feathering
FP	. . Manoeuvrability forces
FPH	. . . Performance (aerodynamics)
FPJ	. . . Speed
FPL	. . . . Rate of ascent
FPN	. . . . Rate of descent
FPP	. . . Take-off
FPR	. . . Landing (aerodynamics)
FT	. Interference
	. . * Aerodynamic influence of one body on another.
G	. Aerodynamic heating
GH	. Downwash
GQ	. Slipstream

Special attributes & elements of flow

[Fluids BS]
[Types of fluids]
[Gases BT]
[Mechanics]
[Pneumatics BTB]
[Special attributes & elements of flow]
. [Slipstream BTGGQ]

BTG GT . Turbulence
GX . Ground effect, ground resonance
JQ . Sources
JS . Sinks
JT . Doublets
L . Jets
LU . . Couette flow
NC . Leaks
NG . Eddies
NN . Circulation
. . * Lift-producing airflow round an
. . aerofoil.
NP . . Supercirculation
P . Boundary layer
. . *Processes*
PH . . . Transition
PK . . . Separation
PL . . . Reattachment
PM . . . Thickening
PN . . . Growth
PP . . . Interference
. . *Types*
PS . . . Wall boundary layer
PT . . . Oscillating boundary layer
Q . . . Other [
. . . . * Add to BTG Q letters DF/L
. . . . following BT; eg Vortex boundary
. . . . layer BTG QDP : Laminar
. . . . boundary layer BTG QI.
Types of gas flow
. *By shape*
RC . . Conical flow
. *By compressibility*
. . * See also Rarefied gas flow BTG UR
RE . . Compressible flow
RG . . Incompressible flow
. *By viscosity*
RJ . . Viscous flow
RJL . . . Slow viscous flow
RJP . . . Creeping viscous flow
RK . . Non-viscous flow
RL . . Free stream flow
RM . . Newtonian flow
RN . . Non-Newtonian flow
RNL . . . Relativistic flow
. *By symmetry*
SE . . Symmetrical flow
SF . . Axisyymetrical flow
. *By equilibrium*
SH . . Equilibrium flow

[Types of fluids]
[Gases BT]
[Mechanics]
[Pneumatics BTB]
[Types of gas flow]
[By equilibrium]
. [Equilibrium flow BTGSH]

BTG SJ . Non-equilibrium flow
By degree of attachment
SM . Separated flow
SN . Unseparated flow
SR . Reattached flow
SS . Mixed flow
By stress
SV . Cavity flow
T . . Shear flow
By development
TQ . Secondary flow
TT . Acceleratd flow
By pressure
. * For compressible & incompressible flow,
. see BTG RE.
U . Isobaric flow
UR . Rarefied gas flow
V . . Molecular flow, Knudsen flow
. . . * Mean free path of the gas molecules
. . . is large compared with dimensions of
. . . containing vessel; so flow rate is
. . . determined more by molecule/wall
. . . collisions than intermolecular collisions
. . . (ie viscosity).
VRL . . . Free molecular flow
W . . Superaerodynamics
. . . * Flow at very low densities.
X . . Slip flow
XP . . . Expansion flow, Prandtl-Meyer
. . . flow
Y . Isothermal flow
By parameter changes
BTH B . Reversible flow
D . Irreversible flow
By presence of velocity potential
L . Potential flow
. . Equipment
L3Y U . . . Electrolytic tanks (potential
. . . flow)
By continuity
S . Steady flow, continuous flow
. . * For Couette flow, see Jets BTG LT.
U . Unsteady flow, transient flow
By degree of mixing
W . Free stream flow
BTI . Laminar flow, streamline flow,
. Poiseuille flow
BTI S . . Superlaminar flow
T . Transitional flow
BTJ . Turbulent flow

[Types of fluids]
[Gases BT]
[Mechanics]
[Pneumatics BTB]
[Types of gas flow]
[By degree of mixing]
. [Turbulent flow BTJ]

	By speed
BTJ R	. Low speed flow
	. . * Mach 0.3 or less.
S	. Subsonic flow
	. . * M 0.8 - M 1.0.
ST	. . With compressibility effects
SV	. . Without compressibility effects
T	. High speed flow
U	. Subcritical flow
V	. Transonic flow
	. . * Partly subsonic, partly supersonic.
W	. Supersonic flo(
	. . * M 1.0 - M 4.99.
X	. Hypersonic flow
	. . * Mach 5 or more.
Y	. . Hypervelocity flow
	Flow defined by bodies in gases
BTK B	. Bodies in gas flow
BBG	. . Control forces
	. *Parts*
D	. . Surfaces
DR	. . Holes
	. *Types of bodies in flow*
F	. . Streamlined bodies
G	. . One-dimensional bodies gas flow
H	. . Two-dimensional bodies gas flow
J	. . . Manifolds
K	. . . Nozzles
L	. . . Conduits, ducts
N	. . . . Channels, passages
P	. . . . Closed conduits
PP	. . . . . Pipes
PT	. . . . . Tubes
Q	. . Three-dimensional bodies gas flow
R	. . . Wedges
S	. . . Bodies of revolution
T	. . . . Cylinders
BTL S	. . . . Boat-tail bodies
T	. . . . Blunt bodies
U	. . . . Bluff bodies
W	. . . . Immersed bodies
	. . . . . * See Liquids BUL W
X	. . . *Aeronautical structures*
	. . . . * See Aeronautical engineering U

[Fluids BS]
[Gases BT]
[Mechanics]
. . . [Flow defined by bodies in gases]
. . . . [Types of bodies in flow]
. . . . . [Three-dimensional bodies gas flow BTKQ]
. . . . . . [Aeronautical structures BTLX]

BTM	*Other energy interactions & forms*
	. * Normal retroactive synthesis with classes from BR
	. is resumed here after its interruption beginning at
	. BTG FC.
	. * Add to BTM letters GH/Q following BR; eg
BTM GH	. Acoustic properties of gases
GHN	. . Noise
GHS	. . . L-Cavitation noise
GHT	. . . Sonic boom
	. . . . * See also Shock waves BTG A
GP	. Thermal properties of gases
	. . * For aerodynamic heating, see flow BTG G.
	. . Thermodynamics
GPA G	. . . Aerothermodynamics
GQ	. . . Heat transfer
GQT	. . . . Thermal transpiration
GY	. Electrical & magnetic properties
HR	. . Electrical discharge, field discharge
	. . (electricity)
HR3 6	. . . Practical & experimental
HR5 V	. . . . Thermionic tubes
HR5 W	. . . . Cold cathode tubes
HRJ	. . . Partial discharge
HRK	. . . High-frequency discharge
HRL	. . . Electrical breakdown
HRL V	. . . . Electron avalanches
HRM	. . . Surface discharge
	. . . . * See also Luminous discharge (gases)
	. . . . BTM LFH R
HRN	. . . . Corona
HRP	. . . . Brush discharge
HRQ	. . . Disruptive discharge
HRR	. . . . Arc discharge
HRR S	. . . . . Flashover
HRS	. . . . . Spark discharge
HRT	. . . . . Glow discharge
HRV	. . . Space charges
KJ	. . Ionization
KJF C	. . . Ionized gases
	. . Electroluminescence
LFH R	. . . Luminous discharge (gases)
M	*Relations with particle & atomic physics*
	. * Add to BTM letters M/Q following B; eg
NP	. Atomic aspects
	Processes & properties special to bulk matter
BTN P	. Change of state
BTO	. . *Types of change by states involved*
	. . . * See note at BRO re citation order.
	. . . *Change of gas to & from liquid*
	. . . . * See Liquids - Types of change BUO E

[Fluids BS]
[Types of fluids]
[Gases BT]
. [Processes & properties special to bulk matter]
. . [Change of state BTNP]
. . . [Types of change by states involved BTO]
. . . . [Change of gas to & from liquid]

. . . . *Change of gas to & from solid*
. . . . . * See Solids - Types of change BVO E
. *Subsystems, parts*
BTP W . . Impurities
BTQ . . Surfaces
BTQ Q . . Films
. *Types of systems*
. . * For types characterized by physical properties, see the property; eg anisotropic gas BT9 DEL.
. . *By simplifying assumptions*
BTR D . . . Perfect gasesa
F . . . Real gases
. . . . * See also Rarefied gas flow BTG UR
K . . *By number of components*
O . . . Mixtures
. . *By number of phases*
BTS N . . . Homogeneous, single phase
T . . . Heterogeneous, multiphase
TV . . . . Binary systems
U . . . . Anomalous
BTT . . . Dispersions, mixed systems
BTT V . . . . Colloids
BTU . . *By state dispersed*
BTU B . . . Plasmas in gases, gaseous plasmas
G . . . Gases in gases
. . . Gases in liquids & liquids in gases
. . . . * See Liquids - States of matter BUU G
. . . Gases in solids & solids in gases
. . . . * See Solids - States of matter BVU G
. *Types of gases*
. . * For gases by chemical composition, see Class C Chemistry; for air, see DSQ (in Class DG Earth sciences).
Liquids
. * See BU (under condensed matter)

[Bulk matter physics BR]
[States of matter BRN]
[Systems by particular state of matter]
[Plasmas & fluids BRU]
. [Fluids BS]
. . [Types of fluids]
. . . [Liquids]

BTX Condensed matter
. * See note at BR. Although sometimes used to cover all bulk matter, its usual meaning now is of liquids and solids taken together.
. * Add to BTX letters A/U following BR as instructed; eg at BRR, with the addition of BTX BF; eg
BTX BB . Energy
BD . . Energy levels
BDK . . . Fermi level
BF . . Energy bands, band structure
. . . * See also Semi-conductors BVI; Dielectrics BVI S
BF7 5 . . . Calculation
BF7 5W . . . . Calculations (energy bands)
. . . . . * Arrange A/Z; eg OPW calculation BTX BE7 5WO.
BF8 B . . . Theory, models
BF8 H . . . . Relativistic models
BF8 T . . . . Free electron approximation
BF8 U . . . . Nearly-free electron approximation
BF8 V . . . . Cellular method
BF8 W . . . . . Supercell cellular method
. . . *Properties*
BFF . . . . Band edge energy
. . . *Particular bands, states*
. . . . * Some of these are special to semi-conductors (BVI) but are given here for convenience.
BFH . . . . Brillouin zone
BFI . . . . Localized electron states
BFJ . . . . Deep levels (energy bands)
BFK . . . . Inner band
BFL . . . . Allowed band
BFM . . . . Forbidden zone
BFN . . . . Occupied state
. . . . . * Usually implies a metal.
BFO . . . . Valence band
BFP . . . . . Empty state (energy bands)
BFQ . . . . . Energy gaps, band gaps
BFQ R . . . . . . Tails (energy band)
BFR . . . . . Barrier layer
BFR S . . . . . . Barrier penetration
BFS . . . . . Depletion layer
BFT . . . . Conduction bands
BFW . . . . *Special states*
. . . . . * Arrange A/Z; eg
BFW T . . . . . Two level bands
BG . Forces
BJ . . Pressure

[Physics B]
[Bulk matter physics BR]
[States of matter BRN]
[Systems by particular state of matter]
[Condensed matter BTX]
[Forces BTXBG]
. [Pressure BTXBJ]

BTX BJQ . . High pressure physics
BJW . External forces
. . Deformation
BK . . . Rheology
. . . . * Deformation and flow of condensed
. . . . matter.
BQS . . Compressibility
CX Dynamics
DN . Rotation
. . Quanta
DN8 N . . . Rotons
. . . . * Quantum of rotational energy
. . . . (analogous to photon).
GE . *Special attributes & elements of flow*
. . * This class has a special expansion in
. . Fluids BU, for use also in gases and
. . liquids. This special provision applies
. . only to a minor degree to condensed
. . matter in general and probably not at all
. . to solids. But in case any of the concepts
. . concerned are required under BTX or
. . BV (Solids) provision is made here for
. . them:
. . * Add to BTX GE letters GE/LW
. . following BU if applicable.
H Electrical & magnetic properties
HU . Conductivity
IP . . Superconductivity (general)
. . . * Most of the literature on this refers to
. . . the solid state; all details are therefore
. . . given there (at BVI P) but may be
. . . used here to qualify truly general
. . . works.
. . . * See also Superfluids BSV
NP Change of state
O . *Types of change, by states involved*
. . * See note at BRO re citation order.
. . * For vaporization, see liquids BUO G
. . and solids BVO G.
Subsystems
Q . Surfaces
QBD . . Surface energy levels
QBF . . Fermi surface
QO . Interfaces
QOB D . . Interface energy levels

[Bulk matter physics BR]
[States of matter BRN]
[Systems by particular state of matter]
[Condensed matter BTX]
[Subsystems]
. [Interfaces BTXQO]
. . [Interface energy levels BTXQOBD]

BU Liquids, liquid physics
. * Add to BU letters A/U following BS (Fluids).
BU3 6 . Practical
U . . Equipment
YL . . . Water tunnels, water tanks
BUA G . Thermodynamics
T . . Transport properties
. Mechanics
BUB . . Hydromechanics, flow of liquids
BUB J . . . Pressure
JSO . . . . Osmotic pressure
. . . . . * For osmosis, see Chemistry C.
JST . . . . Flotation, floating
. . External forces
BUC B . . . Hydroelasticity
. . . . * See also Vibration BUE
C . . . Viscoelasticity
. . . . * For visoelastic solids, see BVC C.
. . Statics
H . . . Hydrostatics
N . . . . Equilibrium
OW . . . . . Relative equilibrium (liquids(
. . Dynamics
X . . . Hydrodynamics
X2M . . . . Mathematics
X2M 9L . . . . . Continuity equations
X2M 9N . . . . . . *By name*
. . . . * Arrange A/Z; eg
X2M 9NE . . . . Euler's equation
. . . . (hydrodynamics)
BUD K . . . Forms of motions
N . . . . Rotational motion
P . . . . . Vortices
S . . . . Periodic motion
BUE . . . . Vibrations & oscillations
. . . . . * For acoustics in liquids, see BUM GH.
BUF . . . . *Waves*
BUF YU . . . . . Longitudinal
BUG A . . . . . Shock waves
. . . *Special attributes & elements of flow*
. . . . * Normal retroactive synthesis from
. . . . BRG H/BRM is interrupted here in
. . . . order to accommodate the large
. . . . vocabulary of liquid flow. It is resumed at
. . . . BUM.
. . . . * Add to BU letters GE/L following BS; eg
ER . . . . Flow regime
GT . . . . Turbulences
P . . . . Boundary layers
. . . *Types of flow*

[States of matter BRN]
[Systems by particular state of matter]
[Condensed matter BTX]
[Liquids BU]
[Mechanics]
. [Dynamics]
. . [Types of flow]

BUI . . . Laminar flow
BUJ . . . Turbulent flow
BUK B . . . Flow determined by bodies
KN . . . . Channel flow
KO . . . . . Sluices
BUL V . . . . Hydrofoils
W . . . . Immersed bodies, submerged
. . . . bodies, floating bodies
. . . . . Density effects
WCK N . . . . . . Buoyancy
BUM *Other energy interactions & forms*
. * Normal retroactive synthesis is resumed here
. after its interruption at BUG ER.
. * Add to BUM letters GF/Q following BR; eg
BUM GH . Acoustics
GP . Thermal properties
H . Electro-magnetic properties
M . *Relations with particle & nuclear*
. *physics*
BUN P . Change of state
R . . Transitions, transformations
BUO . . *Types of change by states involved*
. . . * See note at BRO re citation order.
. . . Change of liquids to gases & gases
. . . to liquids
BUO E . . . . Liquid-vapour transformations
. . . . . * For works considering both directions
. . . . . of change.
G . . . . . Evaporation, vaporization,
. . . . . volatilization, liquids to
. . . . . gases changes
. . . . . . * For vaporization in the narrower
. . . . . . sense of transformation of solid to
. . . . . . gas, see BVO G.
. . . . . . Entropy
GAM . . . . Entropy of fusion (liquid to
. . . . gas)
. . . . . . Pressure
GBJ . . . . Vapour pressure
. . . . . . Latent heat
GNS . . . . Latent heat of vaporization
GP . . . . . . Volatility
GQ . . . . . . Boiling
. . . . Critical point
GQ9 4C . . . . Boiling point
GQN . . . . Nucleate boiling
GQP . . . . Film boiling
GQR . . . . Pool boiling
GR . . . . . . Formation of vapour below
. . . . . . boiling point

[Liquids BU]
[Other energy interactions & forms BUM]
[Change of state BUNP]
. [Types of change by states involved BUO]
. . [Change of liquids to gases & gases to liquids]
. . . [Liquid-vapour transformations BUOE]
. . . . [Evaporation BUOG]
. . . . . [Formation of vapour below boiling point BUOGR]

BUO GS . . . . . Superheating
. . . . . . * Heating of liquid above its normal boiling
. . . . . . point,
GV . . . . . Field evaporation
H . . . . Liquefaction of gas, change of gases to
. . . . liquids
I . . . . . Condensation
. . . . . . Condensation point
I94 C . . . . Dewpoint
IR . . . . . . Filmwise condensation
IS . . . . . . Dropwise condensation
L . . Change of liquids to & from solids
. . . * See Solids - Change of state BVO L
Subsystems of liquids
BUP W . Impurities
BUQ . Liquid surfaces
BUQ BLX . . Surface tension
OU . . Free surfaces
Q . Films
Systems
. * For systems defined by physical properties, see
. property; eg electrically conducting liquids BUM HU.
. *By simplifying assumptions*
BUR D . . Ideal liquids
N . *By number of components*
O . . Mixtures
. *Systems by number of phases*
BUT . . Liquid dispersions
BUU . . . *By state dispersed*
. . . . * See note at BRT re citation order.
BUU B . . . . Plasmas in liquids, liquid plasmas
G . . . . Gases in liquids & liquids in gases
H . . . . . Liquids in gases
J . . . . . Gases in liquids
L . . . . Liquids in solids & solids in liquids
. . . . . * See Solids BVU L
S . . . . Liquids in liquids
Types of liquids
. * For liquids defined by chemical composition (eg
. water) see Class C Chemistry; for liquid crystals, see
. BWU L

Solids

[Bulk matter physics BR]
[States of matter BRN]
[Systems by particular state of matter]
[Condensed matter BTX]
[Liquids BU]
. [Other energy interactions & forms BUM]
. . [Types of liquids]

BUW D	. . . Pure liquids
BV	Solids, rigid bodies
	. * For solid state physics in the narrow sense of the electrical and electronic properties of solids, see BVH.
	. * Add to BV letters A/U following BR; eg
	. Spatial properties
BV9 DEK	. . Isotropic solids
DEL	. . Anisotropic solids
BVB	. Mechanics
BVB F	. . Energy bands
G	. . Forces
K	. . . Deformation
	. . . . *Parameters*
KL	. . . . . Stress-strain relationships
	. . . . . . Techniques
	. . . . Birefringence techniques
KL6 LFM J	. . . . Photoelastic stress analysis
	. . . . * For photoelelastic effect, see BVL FMJ.
KM	. . . . . . Elastic limit
KN	. . . . . . Limit of proportionality
KP	. . . . . . Strength stress-strain
KQ	. . . . Yield point, yield strength
KR	. . . . Rupture point
KV	. . . . High velocity deformation
L	. . . . Stress
	. . . . . * Deforming force per metre.
	. . . . . *Processes & properties*
L92 D	. . . . . . Stress distribution
L92 E	. . . . Concentration (stresses)
LP	. . . . . . Stress raisers
	. . . . * For example, holes, edges.
LQ	. . . . . . Intensity (stress)
LR	. . . . . . Relaxation
LS	. . . . . . Stress waves
LV	. . . . . Stress components
LX	. . . . . . Tension
	. . . . . *Types of stresses*
MD	. . . . . . Internal stress, residual stress
MF	. . . . . . Combined stress
MH	. . . . . . Static stress
ML	. . . . Hydrostatic pressure (stresses)
MM	. . . . . . Dynamic stress
MN	. . . . Cyclic stress
MP	. . . . . . Flow stress
MQ	. . . . . . Vibrational stress
MR	. . . . . . Thermal stress, thermoelastic stress
MT	. . . . . . Bending stress

[Solids BV]
[Mechanics BVB]
[Forces BVBG]
[Deformation BVBK]
[Stress BVBL]
. [Types of stresses]
. . [Bending stress BVBMT]

BVB N	. . Torsional stress, twisting stress
NS	. . . Impact stress
O	. . Normal stress, direct stress
OT	. . . Tensile stress, tensional stress
OU	. . . . Ultimate tensile stress, direct stress
OV	. . . Compressive stress
P	. . Shear stress
Q	. Loads, loading
	. . * Add to BVB letters Q/V following BTB (aerodynamic loads) so far as applicable.
QS	. . Compressibility
	. . . * See also Bulk modulus BVC BG
T	Strain
TL	. Bauschinger effect
TM	. Hooke's law
	. *By stress*
	. . * Add to BVB U letters L/R following BVB; eg
UN	. . Torsion
	. . . * Result of applying twisting force or torque.
UO	. . Normal strain
UP	. . Shear strain
UW	. Warpage
UX	. Aging
VE	. Elastic deformation
	. . * See also elasticity BVC B
VL	. . Electrostriction, electrostricture strain
	. . . * See also piezoelectric effect BVI UC
VM	. . Dimension change (strains)
VN	. . . Longitudinal strain, linear strain
VO	. . . . Elongation, stretching, lengthening (strain)
VP	. . . . Contraction, shrinkage, shortening (strain)
VQ	. . . Area changes (strain)
VR	. . . Volume changes (strain)
VS	. . . . Swelling (strain)
VT	. . . . Flattening
VX	. . . . Expansion (strain)
WP	. Plastic deformation
WQ	. . Superplasticity
WT	. . Deflection
XD	. . Dislocation
XDB KQ	. . . Yield strength
XF	. . . Slip
XG	. . . . Direction of slip
XH	. . . . Slip planes
XJ	. . . Deformation twinning (dislocation)
	. . . . * Rotation of crystal axes to achieve symmetry.

[Solids BV]
[Forces BVBG]
[Deformation BVBK]
. [Strain BVBT]
. . [Plastic deformation BVBWP]
. . . [Dislocation BVBXD]
. . . . [Deformation twinning dislocation BVBXJ]

BVB XL . . . . Edge dislocation
XN . . . . Screw dislocation
YD . . Failure
YF . . . Fatigue
YG . . . . Limit (fatigue strain)
YH . . . . *By environmental effect*
. . . . . * Arrange A/Z; eg
YHA . . . . . Acoustic fatigue
YHV . . . . . Vacuum fatigue
YJ . . . . *By load*
. . . . . * Arrange A/Z; eg
YJC . . . . . Compression load
YJR . . . . . Random load
YK . . . Creep
YLB . . . . Recovery (creep)
. . . . *By stage*
YLD . . . . . Transient creep
YLF . . . . . Steady state creep
YLH . . . . . Tertiary creep
YM . . . Shearing
. . . . * Angular deformation.
YN . . . Tearing
YP . . . Fracture, rupture (fracture)
YQ . . . . Post-yield fracture
YR . . . . Brittle fracture
YS . . . . Spalling
YT . . . . Cleavage, splitting
YU . . . . Cracks (fracture)
YV . . . Buckling
BVC A Friction, tribology
. * See Surfaces BVQ CA
B Elasticity
. * Property of a physical system allowing it to return to
. the original physical state after removal of a stress.
BE . Modulus of elasticity, elastic constants
. . * Ratio of stress to strain.
. . * For elastic limit, see BVB KM.
BF . . Young's modulus, coefficient of elasticity
BG . . Bulk modulus, volume elasticity
. . . * For compressibility (the reciprocal of bulk
. . . modulus) see BVB QS.
BH . . Shear modulus, elasticity of shear
. . modulus, coefficient of rigidity,
. . rigidity modulus
BI . . Poisson's ratio (elasticity)
BJ . . Rupture modulus
BK . . Bending moment
BL . *Elastic properties*
. . * For relaxation, see BVB LR.
BM . . Ductility

[Condensed matter BTX]
[Solids BV]
[Mechanics BVB]
[Forces BVBG]
. [Elasticity BVCB]
. . [Elastic properties BVCBL]
. . . [Ductility BVCBM]

BVC BN . . . Toughness
BO . . . Hysteresis (properties of solids)
BP . . . Instability (elasticity)
BQ . . . Brittleness
BR . . . Resilience
BS . . . Flexibility (elasticity)
BT . . . Extensibility
BV . . . Stiffness
BW . . . Rigidity
. . . . * For rigidity modulus, see BVC BH.
. . *Forms of elasticity*
. . . * For photoelastic effect, see BVL FMJ.
. . . Viscoelasticity
CC . . . . Viscoelastic solids
. . . . . * Show a combination of liquid-like and solid-
. . . . . like behaviour, due in particular to energy
. . . . . storage and dissipation during deformation.
CE . . . . Anelasticity
. . . . . * See also Creep BVB YK; Damping
. . . . . BVC CJ
CG . . . . Thermoelasticity
CJ . Damping (forces)
. . * For internal friction, see BVQ CAK.
CP . Plasticity
CQ . . Photoplasticity
CS . . Superplasticity
CT . Hardness
D . Internal forces
H Statics
S Motion
X Dynamics
. Periodic motion
BVE . . Vibration, oscillation & wave motion
. . . Coupling
BVE RU . . . . Polaritons
BVG E . . . *Special attributes & elements of flow*
. . . . * This class has a special expansion for fluids,
. . . . gases and liquids, achieved by an interruption
. . . . of normal synthesis. The concepts involved
. . . . apply only to a minor degree to condensed
. . . . matter and hardly at all to solids. But in case
. . . . any of the concepts concerned are required
. . . . here, provision is made for them as follows:
. . . . * Add to BVG E letters GE/LW following BU.
H . . . Acoustic properties (solids)
H7H . . . . Visualization & imaging
H7H L . . . . . Chladni figures
. . . . *Interactions with other energy forms*
HJ . . . . . Magnetoacoustics, acoustomagnetics
HK . . . . . . Acoustomagnetoelectric effects

Electrical & magnetic properties of solids

[Solids BV]
[Mechanics BVB]
[Dynamics BVCX]
. . . [Acoustic properties solids BVGH]
. . . . [Interactions with other energy forms]
. . . . . [Magnetoacoustics BVGHJ]
. . . . . . [Acoustomagnetoelectric effects BVGHK]

BVG O . . . . Extremely high frequencies
. . . . . Quantum state
O8N . . . . . . Phonons
P Thermal properties
. *Interactions with other energy forms*
PJ . . Magnetocaloric effects, thermomagnetic
. . effects, magnetothermal effects
. . . * For pyromagnetism, see BVJ UGX.
W . Low temperature solids
X . High temperature solids
Y Electrical & magnetic properties of solids
. * This is the main schedule for these in Class B, for
. the reasons given at BGY (that many of the concepts
. are dependent on bulk matter). An additional factor
. is that many of them (eg circuits) also reflect the
. existence of human artefacts which, strictly speaking,
. should go in Class U/V Technology.
. * The term Solid state physics is sometimes used for
. this class. For works using it thus, use BVH.
. * Locate works here only when they are treated as
. contributions to physics proper, rather than
. technology. In cases of doubt, prefer technology.
BVH . Electromagnetic properties of solids
. . * Electric and magnetic fields in solids and their
. . interaction with electric charges and currents.
. . * For electric fields, see BVH IBH; for magnetic
. . fields, see BVJ BH.
BVH BH . . Electromagnetic field
. . . * See also Electrostatics BVH N
. . . *Properties*
BHJ . . . . Gradient
BHK . . . . Field strength
BHL . . . . Field interaction
BHM . . . . Excitation of field
BHN . . . . Deexcitation of field
I . *Electricity, electrical properties of solids*
IBH . . Electric field
. . . * Usually implies electrostatic field (BVH NBH);
. . . in cases of doubt, prefer latter.
IBH K . . . Intensity, field strength
JC . . Electrical quantities, electrical variables
. . . *Input-output relations*
JG . . . . Gain
JL . . . . Loss
K . . . Charge, electric charge
. . . . * For free electrons, see BNP RLJ; for charge
. . . . carriers, see BVI E.
KL . . . . Electrification
. . . . . * See also Electrostatics BVH NKL
KNT . . . . Attraction

[Solids BV]
[Mechanics BVB]
[Electrical & magnetic properties of solids BVGY]
[Electricity BVHI]
[Electrical quantities BVHJC]
. [Charge BVHK]
. . [Attraction BVHKNT]

BVH KNV . . Repulsion
KO . . Electric moments
KP . . Polarity
KQ . . Electric dipoles
KR . . Positive charge
KS . . Negative charge
KV . . Hypercharge
L . Voltage, potential difference
. . * For electric potential, see BVH NQ.
L92 D . . Distribution of voltage
. . Gain
LJG . . . Voltage gain
LP . . Voltage drop
LQ . . Overvoltage
LR . . Contact potential
LS . . Surface potential
LT . . Low voltage
LU . . Medium voltage
LV . . High voltage
LW . . Very high voltage
LX . Capacitance
MC . Power (electricity)
. . * Rate at which energy is spent.
. . Gain
MCJ G . . . Power gain
MCP . . Power factor, dielectric power factor
MCR . . Dissipation factor
MCT . . Power level
MD . . Input power
ME . . Output power
MG . . Reactive power
MH . . Apparent power
ML . . Load (electricity)
ML9 2D . . . Load distribution
MN . . . Load fluctuation (electricity)
MP . . . Load instability (electricity)
MQ . . . Load sharing
MR . . . Load energy (electricity)
MS . . . No load
N Electrostatics, static electricity
. * Phenomena associated with electric charge at
. rest, as compared with current electricity.
. * For electrodynamics, see BVH O.
. Electrostatic forces
NBG . . Electric flux lines, electric lines of
. . force
NBH . Electrostatic field
. . * See note at BVH IBH.
NBH K . . Electric field strength, electric
. . intensity

Electrical & magnetic properties of solids

[Solids BV]
[Mechanics BVB]
[Electrical & magnetic properties of solids BVGY]
[Electricity BVHI]
[Electrostatics BVHN]
. [Electrostatic field BVHNBH]
. . [Electric field strength BVHNBHK]

	. . *Types of electrostatic fields*
BVH NBH R	. . . Radial electric field
NBH S	. . . Uniform electric field
NBH T	. . . Crossed fields (electricity)
	. Charge
NK	. . Electrostatic charge
NKL	. Electrification
	. . Induction
NKM	. . . Inductive electrification
NKN	. . Frictional electrification, triboelectrification
	. *Special properties*
NN	. . Edge effect
NQ	. . . Electric potential (edge effect)
	. . . . * Energy needed to move a charge; for potential difference, see BVH L.
O	Electrodynamics
	. * Relationship between electric and magnetic forces and their mechanical causes and effects, especially changes in motion due to electric and magnetic fields.
	. * For quantum electrodynamics, see BNG 8M.
OP	. Electromagnetic induction(
	. . * For alternating currents, see BVH Y
OR	. . Inductance(
OS	. . . Self inductance(
OU	. . . Mutual inductance(
P	. Current(
	. . * Rate of flow of charge.
	. . Distribution of current
P92 D	. . . Conduction, transmission
PCM	. . Density
	. . Kinetics
PDE	. . . Electrokinetics
PEA	. . Oscillation
PR	. . Amplification of current
PS	. . Fluctuations of current
PT	. . Rectification
PV	. . Skin effect
R	. . Electrical discharge (general), field discharge (electricity) (general)
	. . . * Removal or reduction of an electrical charge from a body. Most of the literature concerns discharge in gases; see Bulk matter - Gases BTM HR
S	. . Circuits
	. . . * Arrangement of conductors, etc. to form a path for an electric current.
	. . . * For types of circuits, networks, etc., see electrotechnology U

[Solids BV]
[Mechanics BVB]
[Electrical & magnetic properties of solids BVGY]
[Electricity BVHI]
[Electrodynamics BVHO]
[Current BVHP]
[Circuits BVHS]

BVH S76	Measurement
S76 T	. Q factor
	Conditions, influencing factors
S94 P	. Temperature
S97	. Radiation
	Properties
	. * For capacitance, see BVH LX.
	. Inductance
	. . * See Electromagnetic induction BVH OP
SV	. Immittance
	. . * Admittance and impedance together.
SX	. . Admittance
	. . . * Property which permits flow of current; reciprocal of impedance.
SY	. . . Intensity (admittance)
T	. . . Electric displacement
TX	. . . Susceptance
TY	. . . Conductance
	. . . . * Reciprocal of resistance; real part of admittance.
	. . . Conduction
U	. . . . Conductivity, electrically conducting solids
V	. . Impedance
	. . . * The total opposition to a current flow (by its capacitance and inductance as well as its resistance). In particular, the quantity determining the amplitude of the current for a given voltage in an a-c circuit (see BVH Y).
	. . . * For capacitance, see BVH LX.
VFO	. . . Resonance
VV	. . . Reactance (impedance)
VW	. . . . Capacitive reactance
VX	. . . . Inductive reactance
WB	. . . Resistance
WC	. . . . Constant resistance
WD	. . . . Negative resistance
WE	. . . . Contact resistance
WG	. . . . Insulating resistance
	. . . . *In special materials*

Electrical & magnetic properties of solids

[Electrical & magnetic properties of solids BVGY]
[Electrodynamics BVHO]
[Current BVHP]
. [Circuits BVHS]
. . [Properties]
. . . [Immittance BVHSV]
. . . . [Impedance BVHV]
. . . . . [Resistance BVHWB]
. . . . . . [In special materials]

. . . . * For dielectrics, see BVI S
BVH WJ . . . . . Resistivity (impedance)
. . . . . *Types of impedance*
WL . . . . . . Low impedance
WM . . . . . . High impedance
WN . . . . . . Characteristic impedance
WP . . . . . . Constant impedance
WS . . . Saturation (electricity)
WT . . . Drift (electricity)
. *Types of current*
XB . . Normal current, conduction current
. . . * Usually assumed.
XD . . Displacement current
. . . * Not accompanied by motion of current carriers in the dielectric; formulates electric dipoles.
XF . . Forward current
XH . . Reverse current
XJ . . Critical current
XK . . Eddy current, Foucault current
. . . * For electromagnetic induction, see BVH OP.
. . . * See also Skin effect BVH PV
XL . . Light current
XN . . Heavy current
XV . . Direct current
Y . . Alternating current
YSX . . . Admittance
YTU . . . Susceptance
YWB . . . Resistance
YWD . . . Phase (alternating current)
Electrodynamics of special media
. * For vacuum tubes, see B6I BM; for gas discharge tubes, see B6I BS.

[Condensed matter BTX]
[Solids BV]
[Mechanics BVB]
[Electrical & magnetic properties of solids BVGY]
[Electricity BVHI]
[Electrodynamics BVHO]
[Electrodynamics of special media]

BVI Semicounductors
BVI BCX . Electron states, crystal electron state
. . * For magnetoelectric effect, see BVI UJ.
BCX 8B . . Theory
BCX 8R . . . Localized electron theory
BCX 8S . . . Collective electron theory
BCX HN . . Coulomb effect
BCX J . . Magnetoelectric effect
BCX K . . Radiation effect
BCX KFU F . . . Diffusion
BCX KK . . . Recombination radiation
BCX MH . . Photovoltaic effect
BCX MJ . . Photoconductivity
BD . Energy levels
BF . . Energy bands
. . . * Add to BVI BF letters A/Y following BTX BF; eg
BFL . . . Allowed bands
BFM . . . Forbidden zone
BFO . . . Valence bands
BFT . . . Conduction bands
BFW . . . Other bands
. . . . * Arrange A/Z.
D . *Other energy interactions & forms*
. . * Add to BVI D letters D/Q following B if applicable.
DHK . . Charge
. . . * For Charge carriers, use BVI E.
E . Charge carriers
. . *Processes*
EH . . . Generation
EJ . . . Concentration
EK . . . Injection
EL . . . Recombination
EM . . . Doping
EN . . . . Ion implementation
EP . . . Breakdown
ER . . . . Reverse bias charge
ES . . . . . Avalanche
ET . . . . . Tunnelling, tunnel effect
EU . . . . . . Josephson junction, Josephson effect

Semicounductors

[Electrical & magnetic properties of solids BVGY]
[Electrodynamics of special media]
[Semicounductors BVI]
. [Charge carriers BVIE]
. . [Processes]
. . . [Breakdown BVIEP]
. . . . [Reverse bias charge BVIER]
. . . . . [Tunnelling BVIET]
. . . . . . [Josephson junction BVIEU]

BVI EV . . . . . Zener effect
EX . . . . Electron-bound hole pairs
. . *Properties*
. . . Lifetime
F9C I . . . . Bulk lifetime
FAT . . . Transport properties
FAY C . . . . Mobility
FAY D . . . . . Drift mobility
FCL . . . Density
. . *Constituents*
FJ . . . Traps
. . *Types of carriers*
FN . . . Conduction electrons
FP . . . Ionic conduction
FQ . . . Holes (charge carriers)
FR . . . Majority carriers
FS . . . Minority carriers
GB . Semiconductor materials
. . * More details are available in the Technology schedule U
GC . . Binary materials (semiconductors)
GD . . Ternary materials (semiconductors)
GE . . Intrinsic semiconductor materials
GF . . Extrinsic semiconductor materials
GG . . Dopants, impurities (semiconductors)
GH . . . Acceptor impurities
GJ . . . Donor impurities
. . *By substance*
GL . . . Amorphous substances (semiconductors)
GM . . . Elemental substances (semiconductors)
GN . . . Ionic substances (semiconductors)
GP . . . Liquids (semiconductors)
GQ . . . Oxides (semiconductors)
GR . . . Silicon (semiconductors)
GS . . . Germanium (semiconductors)
GT . . . Group 3-5 semiconductors
GU . . . Gallium arsenide semiconductors
GV . . . Group 2-6 semiconductors
GW . . . . Cadmium sulphide semiconductors
GX . . . . Other substances
HB . Semiconductor devices
. . * See Techniques B6I HB
P Superconductors
. * More details are available from the Technology Class U
. * Add to BVI P letters A/G following BVI so far as applicable; eg
P94 PC . Critical temperature

[Mechanics BVB]
[Electrical & magnetic properties of solids BVGY]
[Electricity BVHI]
[Electrodynamics BVHO]
[Electrodynamics of special media]
[Superconductors BVIP]
. [Critical temperature BVIP94PC]

BVI PET . Tunnelling
PEU . . Josephson effect
PS . Superconducting junction devices, junction devices
PU . . Superconducting quantum interference devices, SQUID
PV . . Meissner effect devices
PW . . Superconducting thin film circuits
S Dielectrics, dielectric materials
. * Substances which can sustain an electric field and act as insulators.
. * See also Piezoelectricity BVI UC
SBV E . Elastic deformation
SBV L . . Electrostriction
SFA . *Wave properties*
SFC . . Propagation
SFC G . . . Loss, dissipation
SFC M . . . Attenuation
SFL . . Absorption
SFP . . Polarization
SFP R . . . Relaxation time
SFU P . . Dispersion
SH . Electromagnetic properties
. Strength
SHB HK . . Dielectric strength, disruptive strength (dielectrics)
. . Loss
SHJ L . . . Breakdown (dielectrics)
. *Electrical properties of dielectrics*
SHP . . Current
. . . Displacement
SHT . . . . Dielectric strain, dielectric displacement
SHT S . . . . . Permittivity
. . . . . . * Ratio of electric displacement in a dielectric to the applied electric field strength.
SHT SU . . . . . . Absolute permittivity
SHT T . . . . . . Relative permittivity, dielectric constant (permittivity)
SHT U . . . . Electric susceptibility
SHT V . . . . . Electric flux (dielectrics)
SHT W . . . . . Dielectric hysteresis
. . . . . . * Quantity of electric displacement across a given area of a dielectric.
SL . Optical properties
SLF MJP . . Photoelasticity
SLH . . Electrooptics

[Electrical & magnetic properties of solids BVGY]
[Electricity BVHI]
[Electrodynamics BVHO]
. [Electrodynamics of special media]
. . [Dielectrics BVIS]
. . . [Optical properties BVISL]
. . . . [Electrooptics BVISLH]

. . . *Types of dielectrics*
BVI SM . . . . Ferroelectric materials
Interactions of electrical with other energy forms
. * Electric phenomena affected by other energy forms. For interactions in which electricity is the agent & not the recipient of the action, see the latter; eg electrooptics BRL H.
UB . Mechanical electric effects, electromechanical effects
UC . . Piezoelectricity
. . . * Production of electrical polarization by mechanical stress. May also refer to the reverse (production of mechanical stress by electrical polarization); in this case, use BVB LH.
. . . * For electrostriction, see BVB VL.
. . . Resistance
UCH WB . . . . Piezoresistance
UCH WH . . . . . Elastoresistance
UCR . . . Ferroelectricity
UGH . Acoustoelectric effects
. . * See also electroacoustics BRG HH
UGP . Thermoelectricity, thermoelectric effect
. . *Special effects*
UGQ . . . Peltier effect
UGR . . . Seebeck effect
UGS . . . Thomson effect, Kelvin effect
UGT . . . Pyroelectricity
UGU . . . Electrocaloric effect
UJ . Magnetoelectric effects, galvanomagnetic effects
UJQ . . Hall effect
UJR . . Nernst effect
UJS . . Photoelectromagnetic effects
UJT . . Magnetoresistance, magnetoresistivity
UJU . . . Corbino effect
V . Photoelectric effects
. . * For electron emission in general, see BNP FG
. . * For photoconductivity, see semiconductors BVI BCX M.
. . * For electroluminescence, see BRL FHR.

[Condensed matter BTX]
[Solids BV]
[Mechanics BVB]
[Electrical & magnetic properties of solids BVGY]
[Electricity BVHI]
. [Interactions of electrical with other energy forms]
. . [Photoelectric effects BVIV]

Magnetism
BVJ Magnetically ordered systems, magnetic materials
BVJ 4 . Instrumentation
5V . . Magnetometers
5W . . Magnetic balances
8M . . Quantum properties
8MF . . . Magnons, quantized spin waves
. Potential energy
BBP . . Magnetomotive force, MMF
BH . Magnetic field
BHK . . Field strength, magnetic intensity
. *Properties & processes*
. . Moments
BIM . . . Magnetic moment
. . . . * For dipole moment. see BVJ NR.
. . Stability
BLR . . . Relaxation (magnetism)
. . . . * Returning to equilibrium after disturbance.
FO . . Resonance
. . . * See also Magnetic resonance imaging B7I O
FOH . . . Double resonance
FP . . Polarization, polarity
JE . . Magnetic loss, magnetic leakage
JF . . Attraction (magnetism)
JG . . Repulsion (magnetism)
JM . . Transition temperature (magnetism)
JN . . Curie temperature, Curie law
K . . Magnetic flux, magnetic lines of force
K9D EL . . . Anisotropy
KL . . . Magnetic circuits
KN . . . Magnetic flux density, MFD, magnetic induction (mfd)
KQ . . . Permeability
. . . . * Ratio of mfd & magnetic field strength.
KR . . . . Magnetic constant
KS . . . Susceptibility
KT . . . Reluctance, magnetic resistance
L . . Magnetization
LM . . . Coercive force
LN . . . Magnetic aftereffect
LP . . . Hysteresis (magnetism)
LR . . . Reversal of magnetization, magnetic polarity reversal
LS . . . Demagnetization, demagnetized fields
LT . . . . Transition (demagnetization)
LU . . . Remanence, residual magnetization
LV . . . Spontaneous magnetization
LW . . . . Magnetic transitions

Electrical & magnetic properties of solids

[Mechanics BVB]
[Electrical & magnetic properties of solids BVGY]
[Magnetically ordered systems BVJ]
[Properties & processes]
. [Magnetization BVJL]
. . [Spontaneous magnetization BVJLV]
. . . [Magnetic transitions BVJLW]

BVJ LX . . . Induced magnetization, magnetic
. . . induction, magnetic displacement
LY . . . Saturation
Field components
. * For fields of force, see BVJ BH
N . Magnetic monopoles
. . * Hypothetical particles underlying magnetism.
NP . Poles
NQ . Dipoles
NQB IM . . Dipole moments
NS . Domains
. . * For domain wall and wall energy, see
. . ferromagnetism BVJ S
Types of magnetic fields
PC . Steady state magnetic fields
PE . . Magnetostatics
PF . High magnetic fields
PL . Multipole fields
PN . Crossed magnetic fields
PP . Geomagnetism
. . * See Earth sciences D
PQ . . Magnetic meridians
Forms of magnetism
. * See also materials defined by these forms; eg
. paramagnetic materials BVR JR.
. *By source*
PX . . Electromagnetism (forms of magnetism)
. . . * Produced by an electric current rather than by
. . . a permanent magnet.
. *By atomic and ionic structure*
. . * See also materials defined by these forms; eg
. . paramagnetic materials BVR JR.
Q . . Diamagnetism
QS . . Soft magnetism, soft magnetic materials
. . . * Unable to retain magnetism.
. . . * See also high polymers BVS RB
R . . Paramagnetism, paramagnetic materials
. . . * See also electron spin resonance BNP MKF O
R6 . . . Techniques
. . . . Cooling
R6G T . . . . . Magnetic cooling
RFO . . . Paramagnetic resonance
RS . . . Pauli paramagnetism, free electron
. . . paramagnetism
RU . . . Superparamagnetism
S . . Ferromagnetism
SLM . . . Coercive force
SNS . . . Domains
SNW . . . Walls
SV . . . Antiferromagnetism

[Solids BV]
[Mechanics BVB]
[Electrical & magnetic properties of solids BVGY]
. [Magnetically ordered systems BVJ]
. . [Forms of magnetism]
. . . [By atomic and ionic structure]
. . . . [Ferromagnetism BVJS]
. . . . . [Antiferromagnetism BVJSV]

BVJ SV . . . . . Antiferromagnetism
T . . . . Ferrimagnetism
TV . . . . . Antiferrimagnetism
U . . *Interactions of magnetism with other*
. . *energy forms*
. . . * Interactions in which magnetic properties are
. . . affected by other energy forms. For
. . . interactions in which the magnetism is the
. . . agent rather than the patient, see the material
. . . or phenomenon affected; e.g.
. . . magnetohydrodynamics BRU J; Kerr effect
. . . (optics) BRL JP.
. . . * See also Magnetic cooling BVR JR6 GT
UB . . . Magnetomechanical effects
. . . . Stress
UBL . . . . . Magnetostriction, piezomagnetism
. . . . . . * Stress dependence of magnetic
. . . . . . properties.
UBL R . . . . . . Joule effect
UCB . . . . . Magnetoelasticty
. . . Rotation
UDN . . . . Gyromagnetic effect
UGX . . . Pyromagnetism
BVK . Radiation properties
BVK 73 . . Production techniques
FC . . Transmission
FC4 . . . Instruments
FC4 AX . . . . Waveguides
BVL . Optical properties
. . * The full schedule for these is at BRL, from
. . which details may be drawn; eg
BVL FE . . Spectra
FIL . . Opacity
FM . . Refraction
. . . Birefringence
FML . . . . Photoelastic effect
FN . . Reflection
FP . . Polarization
FR . . Interference
J . . Magneto-optical effect
L . . Luminosity
LFH R . . . Electroluminescence
M . . Colour
NP . . . Colour systems
BVM *Relations with particle physics*
. * Add to BVM letters M/Q following B.

[Bulk matter physics BR]
[States of matter BRN]
[Systems by particular state of matter]
[Condensed matter BTX]
[Solids BV]
[Mechanics BVB]
. [Relations with particle physics BVM]

BVN P	Change of state
	. * See also types of solids defined by number of phases BVS N/BVU.
R	. Phase transformations
S	. . Latent heat
	. *Types of changes in solids, by states involved*
BVO B	. . *To & from plasmas*
E	. . *To & from gases*
F	. . . Sublimation
FNS	. . . . Latent heat of sublimation
G	. . . . Solids to gases (change of state), vaporization (solid to gas)
H	. . . . Gases to solids (change of state)
L	. . *To & from liquids*
M	. . . Solids to liquids (change of state), liquefactions (of solids)
	. . . . * For humidity, see Class D Meteorology
N	. . . . Melting, fusion (melting)
	. . . . . Critical point
N94 C	. . . . . . Melting point
NAM	. . . . . Entropy of fusion (liquid to solid)
P	. . . . . Plastic-liquid phase
PR	. . . . . . Flow point, softening point
PS	. . . . . . Dropping point, droplets
Q	. . . Liquids to solids (change of state), solidification
	. . . . Critical point
Q94 C	. . . . . Eutectic point, reaction point
R	. . . . Freezing
R94 C	. . . . . Freezing point
S	. . . . Icing, congealing
T	. . . . Supercooling
	. . . . . * Slow & continuous cooling of liquid whereby its temperature drops to below normal freezing point (ie, to a metastable state in which solidification is precipitated by very small additions of solid or by the smallest mechanical disturbance).
W	. . . . Crystallization
	. . . . . * See BWO J
	Subsystems of solids
BVP W	. Impurities

[Bulk matter physics BR]
[States of matter BRN]
[Systems by particular state of matter]
[Condensed matter BTX]
[Solids BV]
[Subsystems of solids]
. [Impurities BVPW]

BVQ	. Surfaces
BVQ B	. . Mechanics
BK	. . . Deformation
CA	. . . Tribology, friction (general)
	. . . . * Friction, lubrication & wear of surfaces in relative motion.
	. . . . *Operations*
CAB	. . . . . Lubrication
CAC	. . . . . . Lubricants
CAD	. . . . *By composition*
	. . . . * Arrange A/Z; eg
CAD G	. . . . Graphite lubricants
CAE	. . . . *By state of matter*
CAE T	. . . . Gaseous lubricants
CAE U	. . . . Liquid lubricants
CAE V	. . . . Solid lubricants
CAF	. . . . Self lubrication
CAG	. . . . Boundary lubrication
CAH	. . . . Thin film lubrication
	. . . . *Processes*
	. . . . . * For friction, see general class BVQ CA.
CAJ C	. . . . . Mechanical contact (surfaces)
CAJ D	. . . . . Adhesion
CAK	. . . . . Internal friction
CAL	. . . . . Coulomb friction
CAM	. . . . . Rolling & sliding friction (together)
CAN	. . . . . . Rolling friction
CAP	. . . . . . Sliding friction
CAQ	. . . . . Skid resistance
CAR	. . . . . Wear
CAR R	. . . . . . Resistance to wear
CAS	. . . . . . Abrasion
CAT	. . . . . . Erosion (tribology)
CAV	. . . . . . Ablation
	. . . . * Removal of matter from surface of a moving body by friction with the atmosphere (eg by vaporization).
	. . . *Elements & attributes of surfaces*
NB	. . . . Roughness, smoothness
NC	. . . . Corrugations
NP	. . . . Protuberences
NR	. . . . Notches
NS	. . . . Perforations
NT	. . . . Porosity
O	. . Interfaces
Q	. Films
	Systems of solids

Solids

[Physics B]
[Bulk matter physics BR]
[States of matter BRN]
[Systems by particular state of matter]
[Condensed matter BTX]
[Solids BV]
[Systems of solids]

	* For systems defined by physical properties, see the property; eg isotropic solids BV9 EK.
	* Chemical aspects of the following classes are located in Class C Chemistry.
	By number of components
BVR N	. Single component systems
O	. Mixtures
OP	. . Eutectics
S	. . Solutions
	. . . * See Homogeneous solids BVS NRS
	By number of phases
BVS N	. Homogeneous solids
	. . Physical and chemical properties are the same about every point; may be amorphous or crystalline.
NL	. . Optically homogeneous solids
NRS	. . Solid solutions
	. . . * Homogeneous mixture of two or more solids, forming a homogeneous crystalline phase of several different chemical species.
NRS NR	. . . Transformation
NRS NRS	. . . . Order-disorder transformations
O	. . Amorphous solids (homogeneous solids)
	. . . *Properties*
OP	. . . . Crystallinity of amorphous solids
P	. . . Vitreous solids
PV	. . . . Glasses
Q	. . . Polymers
RB	. . . . High polymers, plastics
RE	. . . . Elastomers
RG	. . . . Thermoplastics
RH	. . . . Thermosetting plastics
RJ	. . . . Fibres (amorphous solids)
T	. Heterogeneous solids
BVT	. . Dispersions
BVT Q	. . . Surfaces
	. . . *Systems*
V	. . . Colloids
	. . . *By states dispersed*

[Condensed matter BTX]
[Solids BV]
[Systems of solids]
[By number of phases]
. [Heterogeneous solids BVST]
. . [Dispersions BVT]
. . . [By states dispersed]

BVU B	. . . . Plasmas in solid, solid state plasmas
G	. . . . Gas-in-solid & solid-in-gas
H	. . . . . Solid-in-gas dispersions
	. . . . . . Mixtures
HRP	. . . . Dusts
HTV	. . . . . . Colloids
HTW	. . . . Smokes
J	. . . . . Gas-in-solid dispersions
L	. . . . Liquid-in-solid & solid in liquid
M	. . . . . Solid-in-liquid dispersions
MRO	. . . . . . Mixtures
MRP	. . . . Suspensions (solid-in-liquid)
MTV	. . . . . . Colloids
MTW	. . . . Semi-solids
MTX	. . . . Gels
MTY	. . . . Sols
N	. . . . . Liquid-in-solid dispersions
S	. . . . Solid-solid systems
	. . . . . * For solid solutions, see homogeneous mixtures BVS NRS.
BW	Crystals, crystallography
	. * Solids in which the physical properties may vary regularly with direction, being the same along all parallel directions.
	. * An alternative (not recommended) is to locate this class in C, with crystal chemistry.
BW2 M	. Mathematical methods
BW3 6	. Practical & experimental crystallography
U	. . Equipment
BW6 9	. . Physical methods in investigation
K	. . . Radiation techniques
KFM	. . . . Refraction techniques
KFM 9BI	. . . . . Refractive index
KFM J	. . . . . Double refraction, birefringence
KFM K	. . . . . . Ordinary rays (techniques)
KFM L	. . . . . . Extraordinary rays (techniques)
KFP	. . . . Polarization
KFP 76	. . . . . Polarimetry
KFQ	. . . . Diffraction techniques
KFQ H	. . . . . Precession method (diffraction techniques)
KFQ J	. . . . . Rotating crystal method (diffraction techniques)
KFQ L	. . . . . Oscillating crystal (diffraction techniques)
KFQ P	. . . . . Powder method (crystal diffraction)
KFQ R	. . . . . De Jong-Boumann method
KFT	. . . . Scatter techniques
	. . . . *By radiation/particle*

[Solids BV]
[Systems of solids]
[Crystals BW]
[Practical & experimental crystallography BW36]
. [Physical methods in investigation BW69]
. . [Radiation techniques BW6K]
. . . [By radiation/particle]

BW6 LX . . . . X-ray techniques
LXF Q . . . . . X-ray diffraction
LXF QS . . . . . . Laue method
LXF QW . . . . . . Weissenberg method
M . . . . Particulate radiation techniques
NP . . . . . Electron techniques
NPF Q . . . . . . Electron diffraction
NW . . . . . Neutron techniques
NWF Q . . . . . . Neutron diffraction
. *By action on phenomenon investigated*
BW7 M . . Spectroscopy
MLX . . . X-ray spectroscopy
. . . . * See also diffraction techniques BW6 KFQ
Processes & properties
. Spatial properties
. . * For isotropic and anisotropic crystals, see
. . BWS Q.
BWA . Physical crystallography
BWB . . Mechanics
BWB H . . . Forces
K . . . . Deformation
. . . . . * For crystal defects, see BWQ D
KV . . . . . High velocity deformation
L . . . . . Stress
T . . . . . Strain
WP . . . . . Plastic deformation
XD . . . . . . Dislocation
YF . . . . . . Fatigue
YK . . . . . . Creep
YP . . . . . . Fracture
YR . . . . Brittleness
YT . . . . Cleavage
BWC B . . . . Elasticity
BWG P . . Thermal properties
Y . . Electric & magnetic properties
BWL . . Optical properties
BWL FL . . . Absorption
FM . . . Refraction
FP . . . Polarization
FQ . . . Diffraction
M . . . Colour
BWM . . Particle physics & crystallography
BWN P . Change of state
. . * For crystallization, see BWO J.
R . . Phase transformation
T . . . Phase equilibrium

[Solids BV]
[Systems of solids]
[Crystals BW]
[Processes & properties]
. [Change of state BWNP]
. . [Phase transformation BWNR]
. . . [Phase equilibrium BWNT]

Special processes/properties in crystallograph
. * Normal retroactive synthesis by BV (as instructed
. at BR) is modified here in order to accommodate the
. special processes and properties of crystals. Normal
. synthesis is resumed at BWR B.
. * Add to BWN X letters A/W following BVO if
. applicable.
BWO G . Growth of crystals, formation of crystals
H . . Nucleation
. . . * Initiation of the growth process.
. . . *Agents*
HR . . . . Nucleating agents
HS . . . Homogeneous nucleation
HT . . . Heterogeneous nucleation
J . . Crystallization
. . . * Slow formation of a crystal from melt or
. . . solution.
. . . *Agents*
JS . . . . Seed crystals
JW . . . Water of crystallization, constitutional
. . . water, water of hydration
JY . . . Polymorphism (crystallization)
L . . Epitaxy
. . . * Unified crystal growth, depositing one layer on
. . . another.
. . . * See also Semiconductors BVI
LR . . . Liquid phase epitaxy
LS . . . Molecular beam epitaxy
M . . Recrystallization
MS . . . Solid state recrystallization
. . . . * See also Mineralogy DIQ
MT . . Crystal pulling
N . . Thick films (crystal growth)
P . . Whisker crystals
Q . . Dendritic growth, dendrites
R . . Crystalline overgrowth

Structural crystallography

[Condensed matter BTX]
[Solids BV]
[Systems of solids]
[Crystals BW]
[Special processes/properties in crystallography]
. [Growth of crystals BWOG]
. . [Crystalline overgrowth BWOR]

BWO T — Structural crystallography, crystalline state
. * For crystal texture, see BWR C.
T2P — . Geometry
. . * Use BWO U.
T9J S — . Symmetry
. . * Use BWP S.
U — . Geometric crystallography
. *Components*
BWP — . . Lattices, space lattices, arrangement (crystals)
. . . Energy
BWP BB — . . . . Lattice energy
BB2 M — . . . . . Mathematics
BB2 M9N — . . . . . . Kapustinskii equation
. . . Dynamics
CX — . . . . Lattice dynamics
. . . . . * Oscillation of atoms in the lattice.
. . . . . * See also phonons BWP E8N
E — . . . Vibration
. . . . *Special quantum properties*
E8N — . . . . . Phonon
. . . . . . * Quantum of lattice vibrational energy in crystal.
E8N FT — . . . . . . Phonon-phonon scattering
S — . . . Symmetry (crystals)
. . . . * Symmetry is a major defining characteristic of types of crystals. For such types, see BWS Q/BWU.
. . . . * See also Cleavage BWB YT
. . . . *Elements*
. . . . . * For space groups, see BWQ L.
T — . . . . . Point groups, symmetry classes
V — . . . . . Axes of symmetry
. . . . . . * See also types of crystals defined by axes, BWS Q
VR — . . . . . . Crystal axis
VS — . . . . . . Crystallographic axis
VT — . . . . . . Screw axis
VW — . . . . . . Rotational axis of symmetry
W — . . . . . Faces
WR — . . . . . . Crystal habit
X — . . . . . Interfacial angles
. . . . . . Measurement
X76 — . . . . Goniometry (crystals)
BWQ D — . . . Defects in lattice, defects (crystals), imperfections (crystals), impurities (crystals)
D36 — . . . . Practical physics
. . . . *Processes/properties*

[Systems of solids]
[Crystals BW]
[Structural crystallography BWOT]
[Components]
[Lattices BWP]
[Defects in lattice BWQD]
. [Processes/properties]

. . * For doping (implantation of impurities), see semiconductors BVI EM.
BWQ D92 D — . . Distribution of imperfections
D92 E — . . . Concentration of imperfections
DR — . . Annealing
DT — . . Interactions between imperfections
. *Types of defects*
. . *By cause*
EG — . . . Thermal diffusion (crystal defects)
EK — . . . Radiation effects (crystal defects)
. . . . *Processes*
EKG G — . . . . . Channelling effect, blocking
. . . . *By type of radiation*
. . . . . * Add to BWQ E letters KJ/Q following B; eg
ELX — . . . . . X-rays (crystal defects)
. . *By origin*
ER — . . . Intrinsic defects
ET — . . . Extrinsic defects, inclusions (crystal defects)
EV — . . . . Voids, bubbles
EW — . . . . Colour centres
. . *By lattice structural characteristics*
. . . * For line defects, see slip planes BWB XH.
F — . . . Non-stoichiometric defects, defect structures
. . . . * Localized misalignments or gaps in lattice.
. . . . * See also Vacancies BWQ GV
G — . . . Point defects
. . . . *Processes*
GS — . . . . . Defect motion
GT — . . . . . Defect cluster
GV — . . . . Vacancies
GW — . . . . Interstitial point defects, grain boundary defects
H — . . . Perfect crystals
. . . . * Without defects.
J — . . . Twin crystals, twinning
Types of lattice structures
L — . Space groups
LS — . . Electron densities (lattice structures)
LU — . . Unit cells
LV — . . . Bravais lattices
M — . . . Primitive crystal lattices

Structural crystallography

[Solids BV]
[Crystals BW]
[Structural crystallography BWOT]
. . . [Types of lattice structures]
. . . . [Space groups BWQL]
. . . . . [Unit cells BWQLU]
. . . . . . [Primitive crystal lattices BWQM]

BWQ N . . . . . . Face-centred cells
P . . . . . . Body-centred cells
Q . . . . . . Side-centred cells
S . . . . . Slip planes, gliding planes
. . . . . . * Planes of molecular weakness allowing
. . . . . . movement along them without actual
. . . . . . fracture.
. . . . . . Dislocations
SBX D . . . . Glides
SBX GH . . . . Axial glides
SBX GL . . . . Diagonal glides
SBX GN . . . . Diamond glides
T . . . . Reciprocal lattices
Subsystems
. * For crystal faces, see BWP W; for impurities,
. see lattice defects BWQ D.
V . Microstructure (crystals), grain structure,
. crystal texture
W . . Crystallites
X . . Grains
. . . * Aggregations of crystallites in which crystal
. . . structure is continuous across each grain.
XOG . . . Growth
. . . . * See also Annealing BWQ DR
XQD . . . Defects
XR . . . Grain boundaries
Y . . Mixed crystals, solid solutions (crystals)
. . . * Arrangement of atoms or molecules of
. . . different species within same lattice.
YNR . . . Transformation
YNR S . . . . Order-disorder transformations
YR . . . Interstitial solid solutions
YS . . . Substitutional solid solutions
BWR B . Surfaces
. . * Normal synthesis is resumed here after its
. . interruption at BWO.
. . * Add to BWR B letters B/V following BVQ if
. . applicable; eg Smoothness BWR BNB.

[Systems by particular state of matter]
[Condensed matter BTX]
[Solids BV]
[Systems of solids]
[Crystals BW]
[Subsystems]
. [Surfaces BWRB]

Systems, types of crystals
. * Types defined by physical properties are
. collocated with the properties unless
. otherwise indicated.
. * Add to BW letters RN/U following BV so
. far as applicable, with the modifications
. indicated below; eg
BWR M . Crystal systems (general)
. . * See also semi-conductors systems BVI
. *By number of phases*
BWS RB . . Plastic crystals
BWU L . . Liquid crystals

. *By positional properties*
BWV D . . Disordered systems, disordered
. . crystalline alloys
. . . * Atoms are displaced from crystal
. . . lattice (eg, by ionizing radiation) to
. . . positions which are not part of the
. . . lattice.
. *By dimensional properties*
F . . Thin crystals
H . . Powder crystals
. *By number of components,*
. *aggregation*
J . . Single crystals
. . . * For twin crystals, see BWQ J.
K . . Polycrystalline crystals
. . . * Aggregates of single interlocking
. . . crystals.
. *By symmetry*
M . . Crystal classes (general)
. . . * If distinguished from crystal systems.
. . . * For point groups, see BWP T.
N . . Body-centred crystals
P . . Face-centred crystals
Q . . Close-packed crystals
. *By crystal form*
SD . . Pedion
SF . . Pinacoid
SH . . Dome
SJ . . Prism (crystal forms)
SL . . Sphenoid
SN . . Bisphenoid
SP . . Pyramid
SQ . . Bipyramid
ST . . Trapezohedron
SV . . Scalenohedron
. *By relation of planes to axes*
V . . Isotropic crystals

Structural crystallography

[Condensed matter BTX]
[Solids BV]
[Systems of solids]
[Crystals BW]
[Systems]
[By relation of planes to axes]
. [Isotropic crystals BWVV]

BWV X . . Cubic crystals, regular crystals,
. . isometric crystals
BWW B . Anisotropic crystals
C . . Uniaxial crystals
D . . . Tetragonal crystals
E . . . Hexagonal crystals
F . . . Trigonal crystals
G . . . . Rhombohedral crystals
J . . Biaxial crystals
K . . . Rhombic crystals, orthorhombic
. . . crystals
L . . . Monoclinic crystals
M . . . Triclinic crystals
By linearity & behaviour
O . Non-linear crystal systems
P . Piezoelectric crystal systems
Q . Electro-optical crystal systems
S . Scintillation crystal systems
By space group symmetry
BWX . Space group symmetry classes
. . * This location, is reserved for libraries
. . wishing to use the detailed system (of
. . some 230 classes) based on the Federov-
. . Schoenthes system.
By bonding
. * For metal bond crystals, see BWY S.
BWY C . Molecular crystals
F . Ionic crystals, electrovalent crystals
. . Coupling
FFR U . . . Polariton
J . Covalent crystals, valence crystals
By specific element or compound
P . Chemical crystallography
. . * An alternative (not recommended) to
. . subordinating this class to the elements
. . & compounds in C Chemistry. If this
. . option is taken, proceed as follows:
Q . . Metallic bond crystals, metal
. . physics
. . . * See also triclinic crystals BWT H
S . . Other chemical substances
. . . * Add to BWY S letters letter
. . . following C (except for metals).
. . . Notation is povisional.

Making effective use of the A/Z index

1 This index covers two classes - AY General science, and B Physics. When a term appears in the index without qualification but with one classmark in AY and the other in B, the two different contexts should be understood.

2 The fundamental rule observed in the index is that an entry word is never qualified by one of its own subclasses from the schedule. So entries appear for

Absorption BFL
Waves BF

but not for

Waves - Absorption BFL

3 Most terms in the index are elementary ones. No attempt has been made to index the huge number of compound classes which the schedules are capable of forming. For example, there is no entry for the compound class Transport processes in fluids (BSA T). Entries appear for both of the keywords present, but they are for the elementary (general) classes whose combination produces the compound class, not fot the latter itself:

Fluids BS
Transport processes BAST

but not for

Fluids - Transport processes BSAT

4 It is assumed that indexers know that the citation order in the physics class cites a state of matter before an energy form (see Introduction to Class B Physics, section 5). So the two components in the example above are combined in the order Fluids - Transport processes as shown by the classmark BSA T.

5 The procedures observed above reflect the best-known and most effective method for economizing the number of index entries made whilst ensuring that all sought terms still appear. This method (which achieves very large economies) is called chain indexing (see sections 5.76 and 6.5 in the General Introduction to BC2).

6 Never classify directly from the A/Z index. Always turn to the schedules, which show each term in its context; the latter often affects the meaning. There may also be special instructions about building classmarks in the class concerned; eg at BSG E Fluids - Elements and types of flows the normal classmark building is interrupted. So classes from BG/BL are added to BSM, not BS; eg Electrical properties of flow is BSM H and not BSH as it would be normally. Classifying from the index would result in a serious error here.

B

C

E

F

Index

G

H

I

J

K

L

N

O

P

Q

R

S

T

U

V

W

X

Y

Z

Printed and bound by CPI Group (UK) Ltd, Croydon, CR0 4YY
13/09/2026
14960860-0001